SILICON-ON-INSULATOR TECHNOLOGY: Materials To VLSI

THE KLUWER INTERNATIONAL SERIES IN ENGINEERING AND COMPUTER SCIENCE

VLSI, COMPUTER ARCHITECTURE AND DIGITAL SIGNAL PROCESSING

Consulting Editor
Jonathan Allen

Latest Titles

BiCMOS Technology and Applications, A. R. Alvarez, Editor
ISBN: 0-7923-9033-4
Analog VLSI Implementation of Neural Systems, C. Mead, M.Ismail (Editors),
ISBN: 0-7923-9040-7
The MIPS-X RISC Microprocessor, P. Chow. ISBN: 0-7923-9045-8
Nonlinear Digital Filters: Principles and Applications, I . Pitas, A.N.
Venetsanopoulos, ISBN: 0-7923-9049-0
Algorithmic and Register-Transfer Level Synthesis: The System Architect's Workbench, D.E. Thomas, E.D. Lagnese, R.A. Walker, J.A. Nestor, J.V. Ragan,
R.L.Blackburn, ISBN: 0-7923-9053-9
VLSI Design for Manufacturing: Yield Enhancement, S.W..Director, W. Maly,
A.J. Strojwas, ISBN: 0-7923-9053-7
Testing and Reliable Design of CMOS Circuits, N.K. Jha, S. Kundu,
ISBN: 0-7923-9056-3
Hierarchical Modeling for VLSI Circuit Testing, D. Bhattacharya, J.P.
Hayes, ISBN: 0-7923-9058-X
Steady-State Methods for Simulating Analog and Microwave Circuits,
K. Kundert, A. Sangiovanni-Vincentelli, J. White,
ISBN: 0-7923-9069-5
Introduction to Analog VLSI Design Automation, M. Ismail, J. Franca,
ISBN: 0-7923-9102-0
Gallium Arsentide Digital Circuits, O. Wing, ISBN: 0-7923-9081-4
Principles of VLSI System Planning, A.M. Dewey ISBN: 0-7923-9102
Mixed-Mode Simulation, R. Saleh, A.R. Newton, ISBN: 0-7923-9107-1
Automatic Programming Applied to VLSI CAD Software: A Case Study,
D. Setliff, R.A. Rutenbar, ISBN: 0-7923-9112-8
Models for Large Integrated Circuits, P. Dewilde, Z.Q. Ning
ISBN: 0-7923-9115-2
Hardware Design and Simulation in VAL/VHDL, L.M. Augustin, D.C..Luckham,
B.A.Gennart, Y.Huh, A.G.Stanculescu ISBN: 0-7923-9087-3
Subband Image Coding, J. Woods, editor, ISBN: 0-7923-9093-8
Low-Noise Wide-Band Amplifiers in Bipolar and CMOTechnologies,Z.Y.Chang,
W.M.C.Sansen, ISBN: 0-7923-9096-2
Iterative Identification and Restoration Images, R. L.Lagendijk, J. Biemond
ISBN: 0-7923-9097-0
VLSI Design of Neural Networks, U. Ramacher, U. Ruckert
ISBN: 0-7923-9127-6
Synchronization Design for Digital Systems, T. H. Meng ISBN: 0-7923-9128-4
Hardware Annealing in Analog VLSI Neurocomputing, B. W. Lee, B. J. Sheu
ISBN: 0-7923-9132-2
Neural Networks and Speech Processing, D. P. Morgan, C.L. Scofield
ISBN: 0-7923-9144-6

SILICON-ON-INSULATOR TECHNOLOGY: Materials To VLSI

by

Jean-Pierre Colinge
IMEC, Belgium

KLUWER ACADEMIC PUBLISHERS
Boston/Dordrecht/London

Distributors for North America:
Kluwer Academic Publishers
101 Philip Drive
Assinippi Park
Norwell, Massachusetts 02061 USA

Distributors for all other countries:
Kluwer Academic Publishers Group
Distribution Centre
Post Office Box 322
3300 AH Dordrecht, THE NETHERLANDS

Library of Congress Cataloging-in-Publication Data
Colinge, Jean-Pierre.
Silicon-on-insulator technology : materials to VLSI / by Jean-Pierre Colinge.
p. cm. — (The Kluwer international series in engineering and computer science ; SECS 132. VLSI, computer architecture, and digital signal processing)
Includes bibliographical references and index.
ISBN 0-7923-9150-0
1. Semiconductors. 2. Silicon-on-insulator technology.
3. Integrated circuits—Very large scale integration—Materials.
I. Title. II. Series: Kluwer international series in engineering and computer science ; SECS 132. III. Series: Kluwer international series in engineering and computer science. VLSI, computer architecture, and digital signal processing.
TK7871.85.C578 1991
621.381'52—dc20 90-29327
CIP

Printed on acid-free paper.

Printed in the United States of America

Contents

Preface

Silicon-on-Insulator (SOI) technology has been around for over a decade, and yet, there is no book describing its multiple facets and possibilities. In ten years of extensive research efforts, SOI technology has made dramatic progress from the first laser recrystallization experiments to CMOS circuits operating at multi-gigahertz frequencies, withstanding high temperatures or surviving several hundred megarads of radiation. At first SOI technology was only considered as a possible replacement for SOS in some niche applications. It has, however, been discovered since then that thin-film SOI MOSFETs have excellent scaling properties which make them extremely attractive for deep-submicron ULSI applications. The commercial availability of SOI substrates, the good fabrication yield obtained in SIMOX 64k SRAMs, and the demonstration of functional three-dimensional integrated circuits are all indicators of a level of maturity reached by Silicon-on-Insulator technology which cannot be questioned any longer. In this Book, we will try to bridge the gap which exists between the specialized SOI literature and the classical textbooks describing bulk device physics, processing and applications.

The SOI community has been extremely productive in the recent years, and a large amount of papers have been published over SOI materials, devices and circuits. It is, unfortunately, not practical to take all these contributions into account, and a selection had to be made in order to present the most significant results of this research effort in a clear and concise manner. Time is, of course, another limitation. Communications published after the summer of 1990 are, therefore, not included in this Book, with a few exceptions.

The material covered by this bulk is divided in eight Chapters, which are summarized below:

CHAPTER 1: INTRODUCTION briefly describes some obvious advantages of SOI technology, such as the absence of latchup in CMOS structures and the reduction of parasitic source and drain capacitances.

CHAPTER 2: SOI MATERIALS lists the different approaches for producing SOI materials. The basic mechanisms behind the fabrication of thin silicon films on an insulating substrate using epitaxy, melting and recrystallization, implantation or bonding are described. This Chapter also addresses the issues of material quality. The application fields for the different materials are described.

CHAPTER 3: SOI MATERIALS CHARACTERIZATION describes different techniques which have been developed to characterize the physical and electrical properties of Silicon-on-Insulator materials. Indeed, while some "universal" characterization techniques such as SIMS and TEM can, of course, be used to assess the quality SOI materials, some techniques have been developed especially to assess the SOI crystal quality and its interface properties, and to measure the film thickness and the lifetime in Silicon-on-Insulator materials.

CHAPTER 4: SOI CMOS TECHNOLOGY deals with the basics of SOI CMOS processing. Thin-film and thicker-film SOI CMOS processes are compared with bulk CMOS processing. Some process steps which are particular to SOI as well as different MOSFET structures are described.

CHAPTER 5: THE SOI MOSFET deals with the physics of the SOI MOSFET. The electrical characteristics (threshold voltage, body effect and output characteristics) of thick- and thin-film MOSFETs are derived and compared to those of bulk devices. The subthreshold slope and the transconductance of SOI transistors are analyzed in detail. The effects caused by the parasitical bipolar transistor are reviewed.

CHAPTER 6: OTHER SOI DEVICES describes other types of SOI devices (double-gate transistors, bipolar transistors, high-voltage devices, JFETs, optical modulators, ...)

CHAPTER 7: THE SOI MOSFET OPERATING IN A HARSH ENVIRONMENT describes the performances of SOI devices operating in a harsh environment (high temperature, radiations, ...)

And, finally, **CHAPTER 8: SOI CIRCUITS** reviews the performances of modern SOI circuits, such as high-speed CMOS, VLSI, rad-hard and three-dimensional integrated circuits.

Acknowledgements

I would like to express my appreciation to many colleagues who have supported my SOI activity over the years, especially when SOI was considered by most people as nothing but an exotic research topic: Drs. M. Lobet, P. Verlinden and F. Van de Wiele from the Université Catholique de Louvain (UCL), Drs. D. Bensahel, G. Bomchil, E. Demoulin, M. Haond and D.P. Vu from the Centre National d'Etude des Télécommunications (CNET), Drs. S.Y. Chiang, C. Drowley, T. Kamins and J. Moll from Hewlett-Packard, and Drs. C. Claeys, G. Declerck, M. Ghannam, R. De Keersmaecker, H. Maes, R. Mertens and R. Van Overstraeten from the Interuniversitair Micro-Elektronica Centrum (IMEC).

I would like to thank the individuals who helped me collecting information, even unpublished, about SOI technology, and/or who reviewed parts of the manuscript: Drs. A.J. Auberton-Hervé, G. Celler, S. Cristoloveanu, A. De Veirman, E. Dupont-Nivet, J.G. Fossum, M. Haond, P.L.F. Hemment, J.L. Leray, J. Margail, P. Mertens, J.C. Sturm, P. Swart, J. Vanhellemont, D.P. Vu, G. Willems and D. Wouters.

Finally, I would especially like to thank Drs. J.C. Alderman and G.E. Davis for proofreading the whole manuscript and making useful suggestions to improve the content of this Book.

Ce livre est dédié
à mon épouse Carine,
à nos parents,
et à notre fils Colin-Pierre

CHAPTER 1 - Introduction

The idea of realizing semiconductor devices in a thin silicon film which is mechanically supported by an insulating substrate has been around for several decades. The first description of the insulated-gate field-effect transistor (IGFET), which later on evolved into the modern silicon metal-oxide-semiconductor field-effect transistor (MOSFET), is found in the historical patent of Lilienfield dating from 1926 [1.1]. This patent depicts a three-terminal device where what is nowadays called the source-to-drain current is controlled by field effect from a gate which is dielectrically insulated from the rest of the device. The piece of semiconductor which constitutes the active part of the device is a thin semiconductor film deposited on an insulator. In a sense, it can thus be said that the first MOSFET was a Semiconductor-on-Insulator (SOI) device. The technology of that time was unfortunately unable to produce a successfully operating Lilienfield device. IGFET technology was then forgotten for a while, completely overshadowed by the enormous success of the bipolar transistor, discovered in 1947 [1.2].

It was only years later, in 1960, that Kahng and Atalla realized the first working MOSFET [1.3], when technology had reached a level of advancement sufficient for the fabrication of good quality gate oxides. The advent of the monolithic integrated circuits gave MOSFET technology an increasingly important role in the world of microelectronics and CMOS technology is currently the driving technology of the whole microelectronics industry.

CMOS integrated circuits are nowadays almost exclusively fabricated on bulk silicon substrates, and this for two well-known reasons: the availability of electronic-grade material produced either by the Czochralski of by the floating-zone technique, and the possibility of growing a good quality oxide on silicon, a thing which is not possible on germanium or on compound semiconductors. Yet, modern MOSFETs made in bulk silicon are far from the ideal structure described by Lilienfield. Bulk MOSFETs are made in silicon wafers having a thickness of approximately 500 micrometers, but only the first micrometer at the top of the wafer is used for transistor fabrication. Interactions between the devices and the substrate gives rise to a range of parasitic effects.

One of these is the parasitic capacitance between diffused sources and drains and the substrate. This capacitance increases with substrate doping, and becomes larger in modern submicron devices where dopant concentration in the substrate is higher than in previous MOS technologies. Source and drain capacitance consists not only in the obvious capacitance of the depletion regions associated with the junctions, but also in the capacitance between the junction and the heavily-doped channel stop located underneath the field oxide. Another parasitic effect found in CMOS devices is called latchup, which consists in the unwanted triggering of a PNPN thyristor structure inherently present in all bulk CMOS structures. Latchup becomes a severe problem in devices with small dimensions, where the gain of the parasitic bipolar devices involved in the parasitic thyristor become large. Of course, some "tricks" have been found to reduce these parasitic components. The area of the source and drain junctions can indeed be minimized by creating local interconnections and placing contacts over the field area, and occurrence of latchup can be reduced by using epitaxial substrates or deep trench isolation. These different techniques, however, necessitate sophisticated processing, which impacts both the cost and the yield of manufacturing.

If a Silicon-on-Insulator (SOI) substrate is used, quasi-ideal devices can be fabricated. The SOI MOSFET contains indeed the necessary three terminals (a source, a drain, and a gate which controls a channel in which current flows from source to drain). The full dielectric isolation of the devices prevents the occurrence of most of the parasitic effects experienced in bulk silicon devices. To illustrate only a few of them, let us look at Figure 1.1 which schematically represents the cross-sections of both a bulk CMOS inverter and a CMOS SOI inverter. The simplicity of the SOI approach is striking. As pointed out earlier, most parasitical effects in bulk MOS devices find their origin in the interactions between the devices and the substrate. Latchup finds its origin in the parasitic PNPN structure of the CMOS inverter represented in Figure 1.1. The latchup path can be symbolized by two bipolar transistors formed by the substrate, the well and the source and drain junctions. Latchup can be triggered by different mechanisms, such as node voltage overshoots, displacement current, junction avalanching and photocurrents. A necessary condition for latchup occurrence is that the current gain of the loop formed by the two bipolar transistors be larger than unity ($\beta>1$)[1.4]. In an SOI CMOS inverter (the silicon film is thin enough for the junctions to reach through to the buried insulator), such a latchup path is ruled out because there is no current path to the substrate, and the lateral PNPN structures contain heavily doped bases (the N^+ and P^+ drains), the heavy doping of which reduces the gain of the bipolar devices to virtually zero.

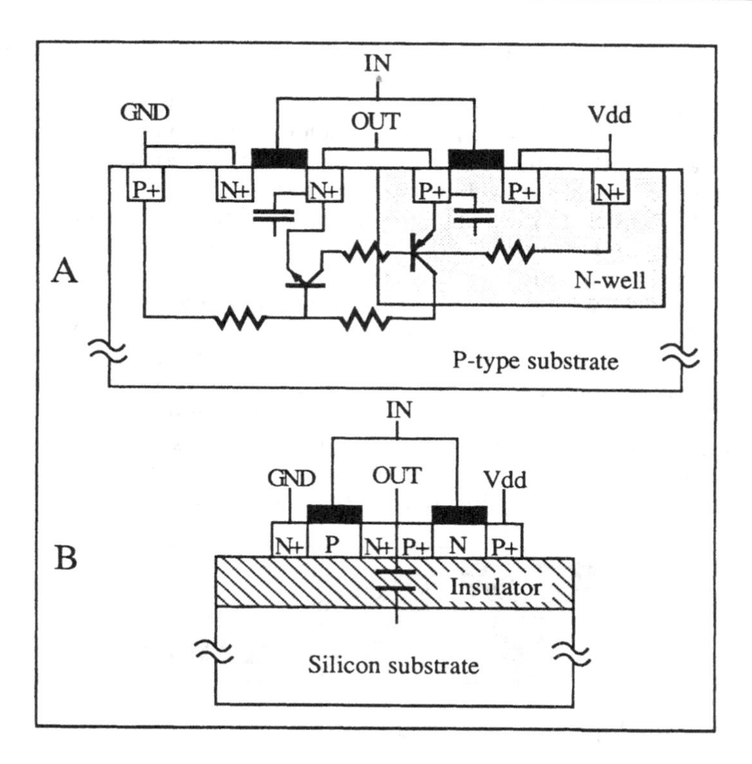

Figure 1.1: Cross section of a bulk CMOS inverter (A) showing a latchup path and an SOI CMOS inverter (B). The drain parasitic capacitances are also presented.

Bulk circuits utilize reverse-biased junctions to isolate devices from one another. Let us consider, for instance, the drain of the n-channel transistor of Figure 1.1. The drain is always positively biased with respect to the substrate (the drain voltage can range between GND and $+V_{DD}$). A depletion capacitance is associated to the drain junction. Its maximum value is reached when the drain voltage is 0 volt and depends on the substrate dopant concentration. The higher this dopant concentration, the higher the capacitance. Modern submicron circuits tend to use higher and higher dopant concentrations. This contributes to increase junction capacitances. In addition, there exists also an important parasitic capacitance between the junctions and the channel stop implant placed underneath the field oxide to prevent surface leakage between bulk devices.

In SOI circuits, on the other hand, the maximum capacitance between the junctions and the substrate is the capacitance of the buried insulator (the capacitance tends towards zero if thick insulators are used, which is the case in SOS technology). This capacitance is proportional to the dielectric constant of the capacitance material. Silicon dioxide, which is widely used as buried insulator, has a dielectric constant (ε_{ox}=3.9 x ε_o) which is three times smaller than that of silicon (ε_{si}=11.7 x ε_o). Therefore, a junction located on a buried oxide layer gives rise to a parasitic capacitance which is three times smaller than that of a bulk junction giving rise to a depletion depth equal to the buried oxide thickness. Buried insulator thickness does not need to scale down as devices with smaller dimensions are produced, and, hence, parasitic capacitances do not increase as technology progresses, contrarily to what happens in bulk devices. In addition, a lightly-doped, p-type silicon wafer can be utilized as mechanical support. In that case, a depletion layer can be created beneath the insulator, which further contributes to reduce the junction-to-substrate capacitances [1.5].

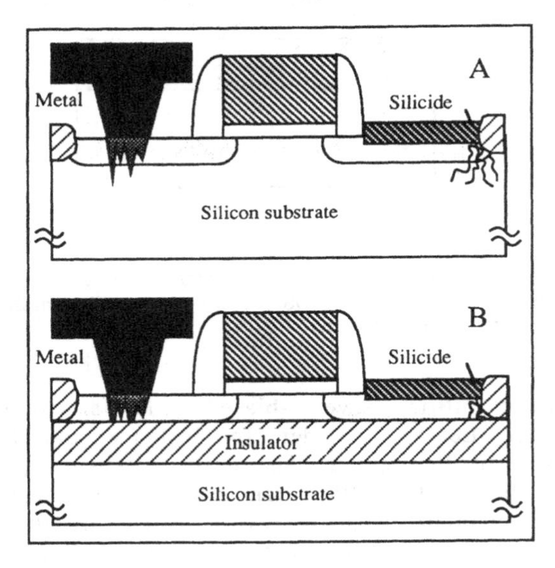

Figure 1.2: Formation contact or silicide on shallow junctions in the case of bulk silicon (A) and thin-film SOI (B).

Silicon-on-Insulator CMOS technology is also attractive because it is simpler than bulk CMOS technology and because it suppresses some yield hazard factors present in bulk CMOS. To illustrate this we will take the example of realizing a shallow junction and making contact to it (Figure 1.2). Forming a shallow (< 100 nm) junction is not an obvious task in bulk CMOS. If a thin (100 nm)SOI substrate is used, on the other hand, the depth of the junction will automatically be equal to the thickness of the silicon film. This somehow relaxes the thermal budget constraints of the process. Contact to a shallow junction can be made using a metal (*e.g.*. tungsten), an alloy (*e.g.* Al:Si or Ti:W) or a metal silicide (*e.g.* $TiSi_2$). In bulk silicon devices, unwanted reactions can sometimes occur between the silicon and the metal or the silicide, such that the metal "punches through" the junction (Figure 1.2.A). This effect is well known in the case of aluminum (aluminum spiking), but can also occur with other metal or silicide systems, especially along field isolation edges. Such a junction punch-through gives rise to uncontrolled leakage currents. If the devices are realized in thin SOI material, the N^+ or P^+ source and drain diffusions extend to the buried insulator (reach-through junctions). In that case, there is no metallurgical junction underneath the metal-silicon contact area, and no leakage will be produced if some uncontrolled metal-silicon reaction occurs (Figure 1.2.B).

The absence of latch-up, the reduced parasitic source and drain capacitances, and the ease of making shallow junctions are merely three obvious examples of the advantages presented by SOI technology over bulk. There are many other properties which allow SOI devices and circuits to exhibit performances superior to those of their bulk counterparts (radiation hardness, improved transconductance and subthreshold slope, ...). These properties will be described in detail in Chapters 5 and 7.

CHAPTER 2 - SOI Materials

Many techniques have been developed for producing a film of single-crystal silicon on top of an insulator. Some of them are based on the epitaxial growth of silicon on either a silicon wafer covered with an insulator (homoepitaxial techniques) or on a crystalline insulator (heteroepitaxial techniques). Other techniques are based on the crystallization of a thin silicon layer from the melt (laser recrystallization, e-beam recrystallization and zone-melting recrystallization). Silicon-on-insulator material can also be produced from a bulk silicon wafer by isolating a thin silicon layer from the substrate through the formation and oxidation of porous silicon (FIPOS) or through the ion beam synthesis of a buried insulator layer (SIMOX, SIMNI and SIMON). Finally, SOI material can be obtained by thinning a silicon wafer bonded to an insulator and a mechanical substrate (wafer bonding). Every approach has its advantages and its pitfalls, and the type of application to which the SOI material is destined dictates the material to be used in each particular case. SIMOX, for instance, seems to be an ideal candidate for VLSI and rad-hard applications, wafer bonding is more adapted to bipolar and power applications, while laser recrystallization is the main contender for the fabrication of 3D integrated circuits. This Chapter will review the different techniques used for producing SOI material.

2.1. Heteroepitaxial techniques

Heteroepitaxial Silicon-on-Insulator materials are obtained by epitaxially growing a silicon film on a single-crystal insulator. Reasonably good epitaxial growth is possible on insulating materials which have lattice parameters sufficiently close to those of single-crystal silicon. Substrates can either be single-crystal bulk material, such as $(0\ 1\ \bar{1}\ 2)$ Al_2O_3 (sapphire) or thin insulating films grown on a silicon substrate (epitaxial CaF_2). Heteroepitaxial growth of a silicon film can never produce a defect-free material by itself if the lattice parameters of the insulator do not match perfectly those of silicon

(Table 2.1.1). Furthermore, the silicon film will never be stress-free if the thermal expansion coefficients of the silicon and the insulator film are not equal.

Material	Crystal Structure	Dielectric constant	Lattice parameter (nm)	Mean thermal expansion coefficient 20-1000°C, (/°C)
Si	Cubic	11.7	0.54301	3.8 E-6
Sapphire (0 1 -1 2)	Rhombo-hedral	9.3	0.4759	9.2 E-6
Cubic Zirconia	Cubic	38	0.5206	11.4 E-6
Spinel	Cubic	8.4	0.808	8.1 E-6
CaF2	Cubic	6.8	0.5464	26.5 E-6

Table 2.1.1: Parameters of the main materials involved in heteroepitaxial SOI technologies [2.1].

Heteroepitaxial silicon-on-insulator films are grown using silane or dichlorosilane at temperatures around 1000°C. All the insulators have thermal expansion coefficients which are 2 to 3 times higher than that of silicon. Therefore, thermal mismatch is the single most important factor determining the physical and electrical properties of heteroepitaxial silicon films grown on bulk insulators. Indeed, the silicon films have a thickness which is typically 1000 times smaller than that of the insulating substrate. While the films are basically stress-free at growth temperature, the important thermal coefficient mismatch results in a compressive stress in the silicon film which reaches $\cong -7 \times 10^9$ dyne/cm^2 in the case of a 0.5 μm SOS film, for instance, while a still higher value is reached at the Si-sapphire interface. Such stresses equal or exceed the yield stress of silicon. Relaxation in the silicon film takes thus place through generation of crystallographic defects such as microtwins, stacking faults and dislocations.

The most important heteroepitaxial SOI techniques will now be briefly reviewed, and their advantages and pitfalls will be outlined.

2.1.1. Silicon-on-sapphire

There is no doubt that silicon-on-sapphire (SOS) is the single most mature of all heteroepitaxial SOI techniques. Until recently, it was the only SOI technology able to produce LSI-VLSI circuits. Several milestones of SOS technology are listed in Table 2.1.2 [2.2].

1963	Idea of SOS (Manasevit and Simpson) [2.3]
1971	Wafers commercially available
1975	1k SRAM (RCA) [2.4]
1976	16-bit microprocessor (HP) [2.5]
1977	4k SRAM (RCA) [2.6]
1978	16-bit, 7K gate microprocessor (Toshiba)
1978	16k CMOS SRAM (RCA) [2.7]
1980	16-bit high-speed microprocessor (Toshiba)
1982	5-inch SOS wafers (Kyocera)
1984	Subnanosecond CMOS gate array (Toshiba) [2.2]
1987	64k CMOS SRAM (Westinghouse) [2.8]
1988	4-bit, 1 GHz flash ADC (Hughes) [2.9]
1988	Thin-film SOS devices [2.10]

Table 2.1.2: Some milestones of SOS technology

The sapphire (α-Al_2O_3) crystals are produced using either the flame-fusion growth technique, Czochralski growth, or edge-defined film-fed growth [2.13]. The two first of these techniques provide sapphire boules which have to be sliced before polishing, while the third technique produces thin rectangular sapphire ribbons which have to be cut into circular wafers later on. After mechanical and chemical polishing, the sapphire wafers receive a final hydrogen etching at 1150°C in an epitaxial reactor, and a silicon film is deposited using pyrolysis of silane at a temperature between 900 and 1000°C. Due to lattice and thermal mismatch, defect density in the films is quite high, especially in very thin films. As the film thickness increases, however, the defect density appears to decrease as a simple power law function of the distance from the Si-Sapphire interface. The main defects present in the as-grown SOS films are: aluminum autodoping from the Al_2O_3 substrate, stacking faults and microtwins. Typical defect densities near the Si-Sapphire interface reach values as high as 10^6 planar faults/cm and 10^9 line defects/cm^2 [2.14]. These account for the low values of resistivity, mobility, and lifetime near the interface. The lowest aluminum autodoping is obtained at deposition rates of 2-2.5 μm/min [2.13], such that aluminum doping concentrations below the SIMS detection limit ($\cong 10^{15}$ cm^{-3}) can be obtained [2.14].

The electron mobility observed in SOS devices is lower than bulk mobility. This is a result of both the high defect density found in as-grown SOS films and the compressive stress measured in the silicon film. Indeed, in the case of (100) SOS, the compressive stress causes the kx and ky ellipsoids to become become more populated with electrons than the kz ellipsoid (which is normal to the silicon surface). As a result, the effective mass of the electrons in the inversion field becomes larger than in bulk silicon [2.2]. Consequently a relatively low channel electron mobility is observed in SOS MOSFETs (≅250-350 $cm^2/V.s$). On the other hand, the effective mass of holes is smaller than in bulk silicon, due to the same compressive stress. The hole surface mobility, which could in principle be higher than in bulk, is, however, affected by the presence of defects, so that the value of mobility for holes in SOS films is comparable to that in bulk silicon. A beneficial consequence of the very low electron mobility at the Si-Sapphire interface is the reduction of the back-channel leakage current. The minority carrier lifetime found in as-grown SOS films is a fraction of a nanosecond. As a result, relatively high junction leakage currents (≅1 pA/μm) are observed [2.1].

Several techniques have been developed to reduce both the defect density and the stress in the SOS films. The first of these is melting and recrystallization of the silicon layer by means of a pulsed laser [2.15]. This technique brings about a 33% increase of the electron field-effect mobility, but it can have detrimental effects on both the leakage current and the threshold voltage uniformity, due to interface state generation.

The Solid-Phase Epitaxy and Regrowth (SPEAR) and the Double Solid-Phase Epitaxy (DSPE) techniques are other more successful methods for improving the crystal quality of SOS films [2.16-18]. These techniques employ the following steps. At first, silicon implantation is used to amorphize the silicon film, with the exception of a thin superficial layer, where the original defect density is lowest. Then a thermal annealing step is used to induce solid-phase regrowth of the amorphized silicon, the top silicon layer acting as a seed. A second silicon implant is the used to amorphize the top of the silicon layer, which is subsequently recrystallized in a solid-phase regrowth step using the bottom of the film as a seed. In the SPEAR process, an additional epitaxy step is performed after solid-phase regrowth. Using such techniques, substantial improvement of the defect density is obtained. Noise in the MOS devices is reduced, and minority carrier lifetime is increased by two to three orders of magnitude, up to 50 ns [2.19]. Typical improvements brought about by the DSPE process are: from 300 to 450 $cm^2/V.s$ for electron mobility, from 185 to 250 $cm^2/V.s$ for hole mobility, from 110 to 92 mV/dec for n-channel subthreshold slope, from 1 to 0.4 pA/μm for the n-channel leakage (V_{DS}=3 V), and 47 to 63 mA/mm for the drive current of nMOS devices (L_{eff}=1.4 μm, V_G=V_{DS}=3V) [2.18]. More recently, thin (0.1-0.2 μm) high-quality SOS films have been produced, and MOSFETs with excellent performances have been fabricated in these films [2.10]. Field-effect mobilities of 800 and 250 $cm^2/V.s$ have been reported for electrons and holes, respectively, in devices made in these thin SOS films.

Silicon-on-sapphire is not only a subset of SOI technology but it is also a mature technology in itself. We will not further describe SOS technology in this Book, in order to focus our attention on new emerging SOI technologies. The reader who is interested in more information about SOS technology can always refer to the excellent review article written by A.C. Ipri in Reference [2.11].

2.1.2. Cubic zirconia

Yttria-stabilized cubic zirconia $[(Y_2O_3)_m \cdot (ZrO_2)_{1-m}]$ has recently attracted considerable interest as a substrate for silicon epitaxy [2.20]. Indeed, zirconia is a superionic oxygen conductor at high temperature. This means that, while being an excellent insulator at room temperature ($\rho > 10^{13}\ \Omega$.cm), cubic zirconia allows very rapid transport of oxygen at high temperatures. This unique property has been used to grow an SiO_2 layer at the silicon-zirconia interface (*i.e.* to oxidize the most defective part of the silicon film) by the transport of oxygen through a 500 µm-thick zirconia substrate [2.1, 2.21]. The growth of a 160 nm-thick film at the interface necessitates only 100 min at 925°C in pyrogenic steam. As-deposited, (100) oriented silicon films grown on cubic zirconia have substantially higher crystal quality than state-of-the art SOS, as indicated by the following RBS surface channeling yields: χ_o=0.048 in Si on cubic zirconia (SOZ), 0.12 in SOS, and 0.034 in bulk silicon [2.1].

2.1.3. Silicon-on-spinel

Spinel $[(MgO)_m \cdot (Al_2O_3)_{1-m}]$ can be used as a bulk insulator material or can be grown epitaxially on a silicon substrate at a temperature between 900 and 1000°C [2.22, 2.23]. Stress-free 0.6 µm silicon-on-spinel films have been grown, but the properties of MOSFETs made in this material are inferior to those of devices made in SOS films [2.23]. Better device performance is obtained when much thicker 3-40 µm silicon-on-spinel films are used [2.22]. Again as with SOZ, oxygen can diffuse at high temperature through thin spinel films. Using this property, an Si/spinel/SiO_2/Si structure has been produced. [2.24].

2.1.4. Epitaxial calcium fluoride

Like spinel, calcium fluoride (CaF_2) can be grown epitaxially on silicon. Fluoride mixtures can also be formed and their lattice parameters can be matched to those of most semiconductors. For example, $(CaF_2)_{0.55} \cdot (CdF_2)_{0.45}$ has the same lattice parameters as silicon at room temperature, and $(CaF_2)_{0.42} \cdot (SrF_2)_{0.58}$ is matched to germanium. Unfortunately, the thermal expansion coefficient of these fluorides is quite different from that of silicon, and lattice match cannot be maintained over any appreciable temperature range. Silicon can, in turn, be grown on CaF_2 using MBE or e-gun evaporation at a temperature of approximately 800°C [2.25, 2.26]. As in the case of Si films on epitaxial spinel, Si/CaF_2/Si films are essentially stress-free, which can be readily understood by noticing that the mechanical support is a silicon wafer, which, of course, has the same thermal expansion coefficient as the top silicon film. MOSFETs have been made in Si/CaF_2/Si material and exhibit surface electron and hole mobilities of 570 and 240 $cm^2/V.s$, respectively [2.27].

2.1.5. Other heteroepitaxial SOI materials

Cubic boron phosphide (BP) and rhombohedral $B_{13}P_2$ have also been epitaxially grown on silicon, and silicon has subsequently been grown on these materials [2.28, 2.29]. BP has a lattice parameter of 0.453 nm and a thermal expansion coefficient of $(4\text{-}6.2)\text{x}10^{-6}$/C in the temperature range 127-527°C. BP has a bandgap of only 2 eV, but it can be obtained in a semi-insulating form with a resistivity of 10^{12} $\Omega.cm$ upon a thermal annealing step at 1050°C, which also happens to be the temperature used for epitaxially growing silicon on BP. P-channel MOSFETs with near-bulk mobilities have been fabricated in Si/BP/Si films. A major problem in using BP as an insulator is the narrow window of temperatures which can be used during device processing, due to diffusion of either boron or phosphorus into the silicon film. For completeness, it can also be mentioned that silicon has been grown epitaxially on a wide variety of insulators, such as AlN [2.30], cubic β-SiC [2.31], BeO [2.32], NaCl [2.33], ...

2.1.6. Problems of heteroepitaxial SOI

In addition to the stress in the films and the high crystalline defect density inherent to heteroepitaxy, heteroepitaxial SOI materials suffer from a basic problem: they cannot be processed in standard silicon IC fabrication lines because of contamination problems. It is, indeed, highly undesirable to introduce materials such as sapphire, zirconia, calcium fluoride or boron phosphide in the ovens of a bulk silicon clean room. A second problem is more specific to the use of bulk insulating substrates such as sapphire. These substrates are extremely brittle, and SOS wafer breakage has been an important factor limiting the use of this technology. In addition, SOS, zirconia and bulk spinel substrates are transparent. This poses problems for using commercial steppers for lithography, particle counters, and equipments for measuring the thickness of oxides,... grown or deposited on SOS.

2.2. Laser recrystallization

MOS transistors can be fabricated in a layer of polysilicon deposited on an oxidized silicon wafer [2.34-35], but the presence of grain boundaries brings about low surface mobility values (≅10 $cm^2/V.s$) and high threshold voltages (several volts). Grain boundaries contain silicon dangling bonds giving rise to a high density of surface states (several 10^{12} cm^{-2}) which must be filled with channel carriers before threshold voltage is reached. Above threshold, once the traps are filled, the grain boundaries generate potential barriers which have to be overcome by the channel carriers flowing from source to drain. This gives rise to the low values of mobility observed in polysilicon devices [2.36]. Mobility can be improved and more practical threshold voltage values can be reached if the silicon dangling bonds in the grain boundaries are passivated. This can be performed by exposing the wafers to a hydrogen plasma, during which atomic hydrogen penetrates in the grain boundaries and saturates (passivates) the dangling bonds [2.37]. This treatment can improve the drive current of polysilicon devices by a factor of 10, due to both an increase of mobility and a reduction of threshold voltage. Hydrogen passivation also significantly improves the leakage current of the devices [2.38]. Such devices can find application as active polysilicon loads, and 64k SRAMs with p-channel polysilicon loads have been demonstrated [2.39]. High-performance IC applications, however, require much better device properties, and grain boundaries must be eliminated from the silicon film. This is the goal of all the recrystallization techniques which will be described next.

2.2.1. Types of lasers used

Experiments of laser recrystallization of polycrystalline silicon have been carried out with pulsed lasers (ruby lasers and Nd:YAG lasers) [2.40] at a time where pulsed laser annealing was popular for dopant impurity activation [2.41]. The technique, however, was rapidly abandoned for the fabrication of device-worthy SOI layers because of its lack of controllability and because of the extreme difficulty of growing large silicon grains. Continuous-wave (cw) lasers have proven to be much more effective for producing SOI films. Both cw Argon and CO_2 lasers have been used. Silicon is transparent at the 10.6 µm wavelength produced by CO_2 lasers. Therefore, silicon films cannot be directly heated by such lasers. SiO_2, on the other hand, absorbs the 10.6 µm wavelength with a penetration depth of ≅ 10 µm. Polysilicon films deposited on SiO_2 (quartz or an oxidized silicon wafer) or covered by an SiO_2 cap can thus be melted through "indirect" heating produced by CO_2 laser irradiation [2.42]. CO_2 lasers have the advantage of a high output power (> 100 W), such that wide elliptical beams can be produced, but manipulation of the beam is difficult because the CO_2 infrared radiation is invisible to the human eye. This factor and the fact that the polysilicon layer has to be surrounded by relatively thick SiO_2 layers (to efficiently absorb the 10.6 μm wavelength) have contributed to limit the use CO_2 lasers in SOI materials fabrication. CO_2 lasers are also unpractical for 3D applications, where heating of only the top silicon layer is needed. Continuous-wave (cw) argon lasers, on the other hand, emit two main spectral lines at 488.0 and 514.5 nm (blue and green), and can reach an output power of 25 W when operated in the multiline mode. These wavelengths are well absorbed by silicon. In addition to this, the reflectivity of silicon increases abruptly once melting is reached. This effect is very convenient since it acts as negative feedback on the power absorption and prevents the silicon from overheating above melting point. This "self-limiting" absorption increases process flexibility during recrystallization with an Ar^+ laser (the effect is opposite when a CO_2 laser is used, since solid silicon is transparent at the 10.6 μm wavelength, while molten silicon absorbs CO_2 laser energy through its free carriers). As a consequence of these advantages, the vast majority of SOI-producing experiments based on laser recrystallization have been carried out using Ar^+ lasers. The laser beam is focussed on the sample by means of an achromatic lens (or a combination of lenses) into a circular or, more often, an elliptical spot. Scanning of the beam is achieved through the motion of galvanometer-driven mirrors. The size of the molten zone and the texture of the recrystallized silicon depend on parameters such as laser power, laser intensity profile, substrate preheating temperature (the wafer sits on a heated vacuum chuck), and scanning speed [2.43]. Typical recrystallization conditions of a 500 nm-thick LPCVD polysilicon film deposited on a 1 μm thermal oxide grown on a silicon wafer are: spot size of 50-150 μm (defined as the laser spot diameter at $1/e^2$ intensity, TEM_{00} mode), power of 10-15 watts, scanning speed of 5-50 cm/sec, and substrate heating at 300-600°C. Several sample structures have also been studied, all aiming at improving the crystal quality of the recrystallized silicon film, eliminating grain boundaries and reducing surface roughness. Some of them will be described next, all for the case of recrystallization with an argon laser.

2.2.2. Seeding

Polysilicon films recrystallized on an amorphous SiO_2 substrate have have basically a random crystal orientation. X-ray diffraction studies of polysilicon recrystallized with a gaussian laser profile indicate the presence of crystallites having (111), (220), (311), (400), (331), (110), and (100) orientations [2.44]. This is clearly unacceptable for device fabrication, since different crystal orientations will result in different gate oxide growth rates.

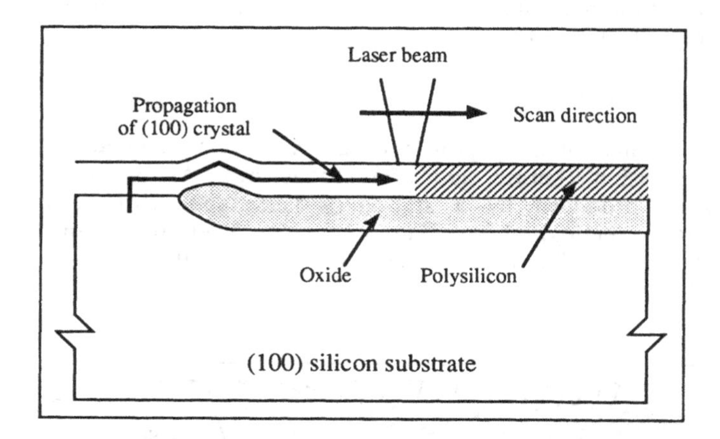

Figure 2.2.1: Principle of the lateral seeding process.

Ideally, one wants a uniform (100) orientation for all crystallites. From there comes the idea of opening a window (seeding area) in the insulator to allow contact between the silicon substrate and the polysilicon layer. Upon melting and recrystallization, lateral epitaxy can take place and the recrystallized silicon will have a uniform (100) orientation (Figure 2.2.1) [2.45]. Unfortunately, the (100) crystal orientation can only be dragged less than 100 μm away from the seeding window, at which distance defects appear which cause loss of the (100) orientation.

2.2.3. Encapsulation

Because of some surface tension and de-wetting effects, polysilicon films have a tendency to "bead up" (form droplets) on SiO_2 upon melting if the laser power is too high. A less dramatic mass transport is observed at optimum laser power, but the flatness of the original polysilicon film is lost after laser melting and recrystallization [2.46, pp. 287 and 288]. In order to solve this problem, a capping layer of SiO_2 and/or Si_3N_4 can be used as a "cast" to improve wetting of the molten silicon and to prevent silicon delamination from the underlying insulator. Silicon dioxide caps are too soft for preventing surface waviness of the recrystallized silicon film [2.47]. On the other hand, nitrogen dissolves in the molten silicon if a Si_3N_4 cap is used. This nitrogen migrates towards the Si-buried insulator interface upon further processing and creates a high density of interface states at this interface [2.48-49]. The best results are obtained using a cap combining both silicon dioxide and silicon nitride. A surface waviness of at least 20 nm seems, however, unavoidable unless a planarization step is used after recrystallization [2.50].

2.2.4. Beam shaping

The normal intensity profile of an Ar^+ laser operated in the TEM_{00} mode is gaussian (before and after focussing). As a consequence, the molten zone produced by a stationary laser spot will be circular (or elliptical if a semicylindrical lens is used). When the beam is scanned across the sample, the molten zone will evolve with time as indicated in Figure 2.2.2.

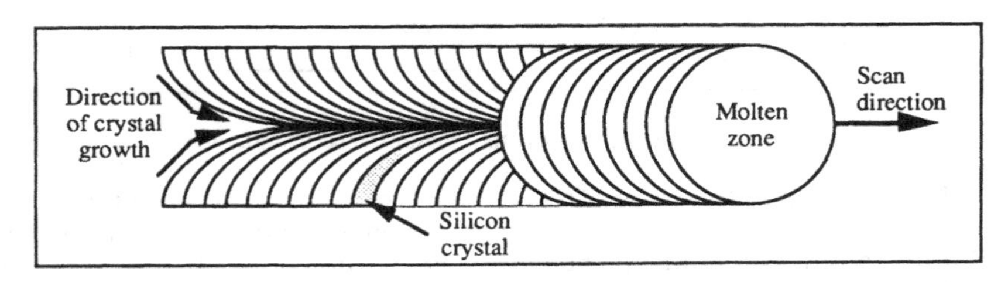

Figure 2.2.2: Chevron recrystallization pattern produced by scanning a gaussian laser beam.

The recrystallization of the quenched silicon proceeds along the thermal gradients which are perpendicular to the trailing edge of the molten zone. As a consequence, crystals grow from the edges of the scanned line towards its center, and in the direction of the scan. The resulting crystal pattern [2.51] is called "chevron pattern" (Figure 2.2.2). The silicon crystals are elongated, with a width of a few micrometers and a length of 10-20 micrometers. The grains are separated by grain boundaries which have detrimental effect on device properties.

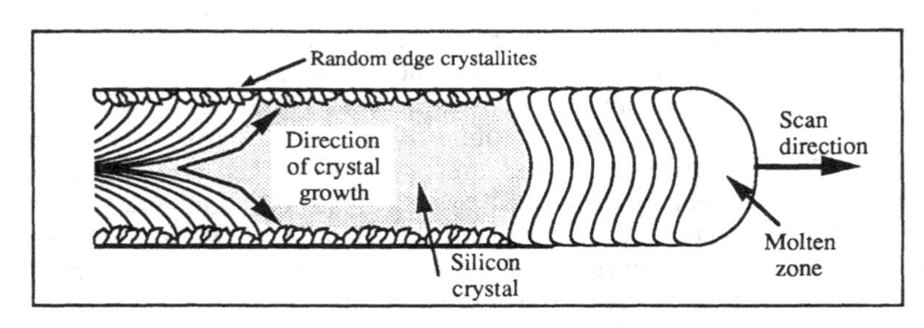

Figure 2.2.3: Growth of a large SOI crystal using a shaped laser beam.

Growth of large SOI crystals can be obtained if the trailing edge of the molten zone is concave. This can be achieved by masking part of the beam (Figure 2.2.3) [2.52]. In this approach, however, the available laser power is reduced. More efficient beam shaping can be achieved by merging different laser modes (doughnut-shaped beam) [2.53] or by recombining a split laser beam [2.54].

In order to obtain not only a single large crystal, but a large single-crystal area, the laser beam has to be raster-scanned on the wafer with some overlap between the scans. Unfortunately, small random crystallites arise at the edges of the large crystals, which precludes the formation of large single-crystal areas, and grain boundaries are formed between the single-crystal stripes.

The location of these grain boundaries depends on the scanning parameters and the stability of the beam. In other words, from macroscopic point of view, the location of the boundaries is quasi-random, and the yield of large circuits made in this material will be zero. A solution to this problem is to use stripes of an antireflecting (AR) material (SiO_2 and/or Si_3N_4) to obtain the photolithographically-controlled beam shaping of an otherwise gaussian beam [2.55-57].

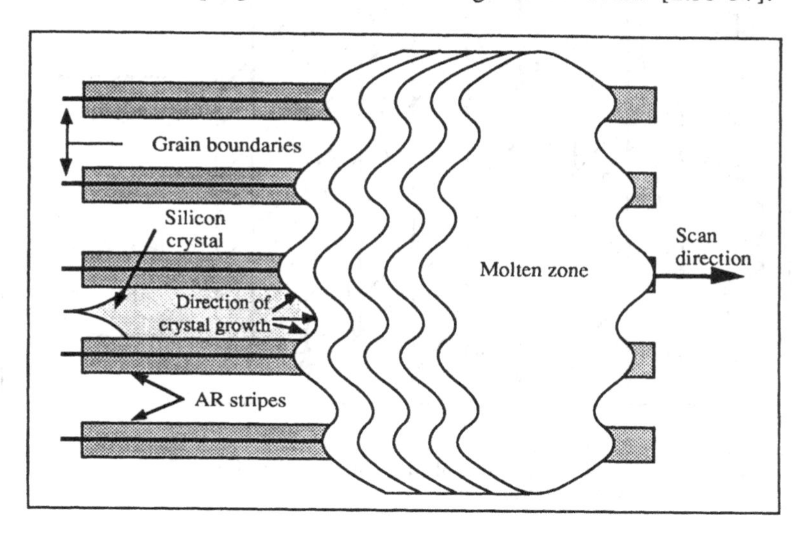

Figure 2.2.4: Recrystallization using antireflection stripes.

This technique is called selective annealing because more energy is selectively deposited on the silicon covered by AR material. It permits one to grow large adjacent crystals with straight grain boundaries, the location of which is controlled by a lithography step (Figure 2.2.4).

Although this technique imposes constraints to circuit design, it allows for placing film defects outside the active area of transistors. The technique can be used with a laser scan parallel, slanted or perpendicular to the antireflection stripes (AR stripes).

Unfortunately, a rotation of the crystal orientation is observed, and the (100) orientation cannot be kept for more that ≅200 μm from the seeding window [2.58]. Defects then appear in the crystals (stacking faults, microtwins or even grain boundaries) until a (110) orientation is reached [2.58]. This rotation can be minimized by forcing the solidification front (the interface between molten and recrystallizing silicon) to be coincident with the [111] facets of the re-solidifying silicon. This can be achieved by using AR stripes parallel to the (150) direction and a laser scan parallel to the (100) direction (Figure 2.2.5). Using this technique, chip-wide (several mm x several mm) defect-free, (100)-oriented single-crystal areas have been produced [2.59].

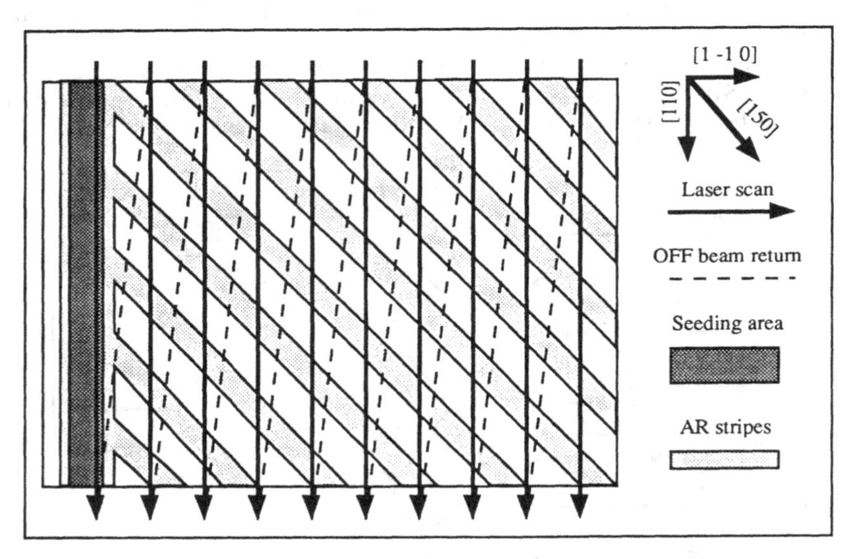

Figure 2.2.5: Recrystallization of a large SOI area using selective annealing, seeding and control of the orientation of the crystal facets in the solidification front.

Recrystallization techniques based on a patterned antireflection coating are primarily used in Japan for the fabrication of three-dimensional integrated circuits (3-D circuits) with up to 4 active device layers [2.60-62].

2.2.5. Silicon film patterning

It is worth mentioning that several other techniques have been developed for producing device-worthy silicon films: periodic seeding windows [2.63], recrystallization of patterned silicon islands [2.64-67] and defect filtering through constrictions of the silicon islands (Figure 2.2.6) [2.68] have been used. These techniques are, however, imposing severe (and often unacceptable) constraints to the circuit layout since the size and shape of the active areas is fixed by laser recrystallization constraints rather than by circuit design considerations.

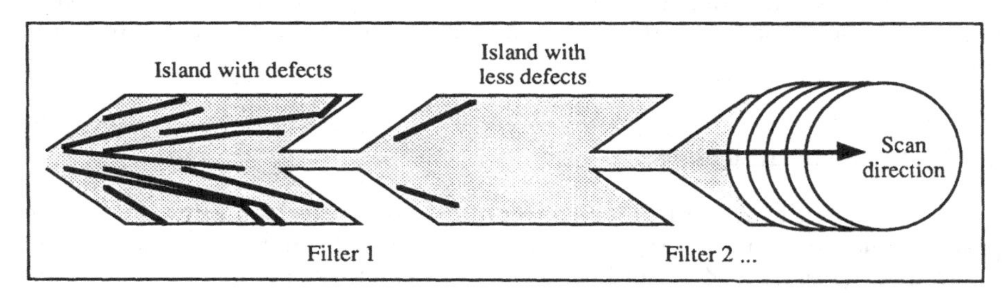

Figure 2.2.6: Principle of defect filtering through constrictions between silicon islands.

While recrystallization of very thin, continuous silicon films has not been reported as yet, recrystallization of very thin (< 100 nm) silicon islands is possible and opens the door to thin-film device 3D IC technology [2.69].

It is also worth noting that several attempts have been made to create an "artificial seed", having a (100) orientation, in the patterned silicon film itself, without the need for opening a seeding window to the substrate [2.70-2.71].

2.3. E-beam recrystallization

The recrystallization of a polysilicon film on an insulator using an electron beam is in many respects very similar to the recrystallization using a cw laser. Similar seeding techniques are used, and an SiO_2 (or Si_3N_4) encapsulation layer is used to prevent the melted silicon from de-wetting [2.72]. The use of an electron beam for recrystallizing SOI layers has some potential advantages over laser recrystallization since the scanning of the beam can be controlled by electrostatic deflection, which is far more flexible than the galvanometric deflection of mirrors. Indeed, the oscillation frequency of a laser beam scanned using galvanometer-driven mirrors is limited to a few hundreds hertz, while e-beam scan frequencies of 50 MHz have been utilized [2.73]. The absorption of the energy deposited by the electron beam is almost the same in most materials, such that the energy absorption in a sample is quite independent of crystalline state and optical reflectivity of the different materials composing it. This improves the uniformity of the recrystallization of silicon deposited over an uneven substrate, but precludes the use of a patterned antireflection coating. Structures with tungsten stripes have, however, be proposed to achieve differential absorption. E-beam recrystallization must be carried out in a vacuum. This presents the disadvantage of precluding the use of a heated vacuum chuck for preheating the substrate, as it is the case in laser recrystallization.

2.3.1. Scanning techniques

The flexibility of electron optics and electrostatic deflection permits one to explore different types of scanning configurations. A first technique (Figure 2.3.1.A) involves the simple raster scan of the focussed electron beam on the sample. This technique produces a chevron recrystallization pattern with grain size up to 20 μm, a texture which looks very much like the one which is obtained by scanning with a gaussian laser spot [2.74]. A second technique makes use of a focussed linear e-beam source (Figure 2.3.1.C) [2.75]. This technique allows for the rapid recrystallization of large areas of SOI, but the control of the uniformity of the beam intensity is rather difficult.

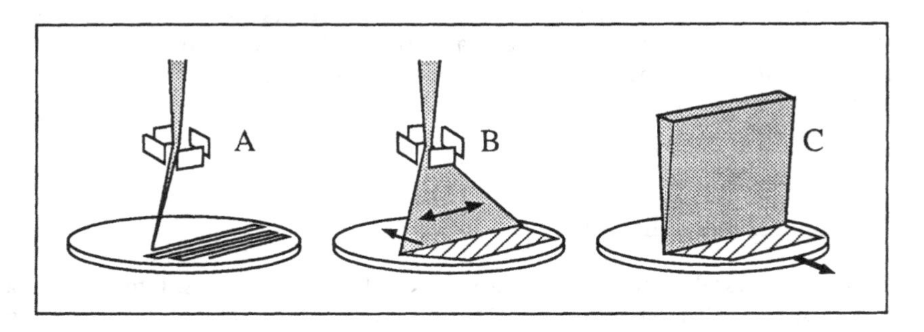

Figure 2.3.1: Beam configurations for e-beam recrystallization: (A) scanned spot, (B) synthesized line source, (C) focussed line source.

The most flexible and powerful e-beam recrystallization technique consists in the synthesis of a pseudo-linear source through rapid scanning of a focussed beam (Figure 2.3.1.B). A continuous, linear molten zone can be created in the silicon film if the period (1/frequency) of the scan is smaller than the thermal constant of the SOI system. This is achieved if a frequency over 2 MHz is used [2.73]. If a sinusoidal scanning is used, the position of the beam is given by $y(t) = W \sin(2\pi ft)$, and the deposited energy, which is proportional to the dwell time at the position y, is significantly higher at the edges of the scanned line (y=W) than at its center (y=0). This can be compensated for by modulating the amplitude of the rapid sinusoidal scan, such that an uniform energy deposition profile can be obtained across the entire scanned line (Figure 2.3.2).

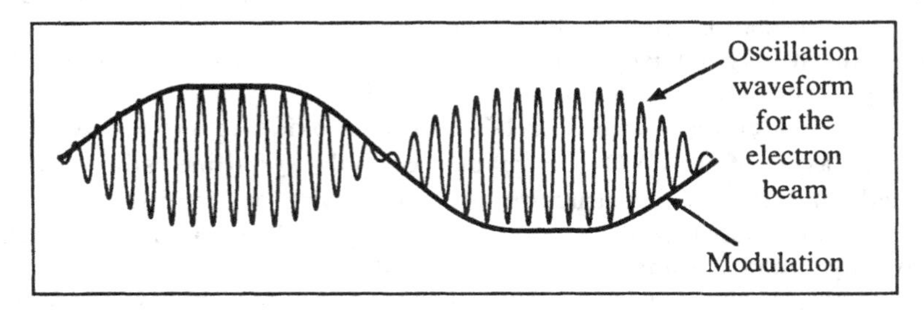

Figure 2.3.2: Electron beam oscillation waveform used to synthesize a uniform-intensity line source [2.73].

2.3.2. Seeding and wafer heating

Two seeding techniques have been proved to be successful when used with e-beam recrystallization. The first one is similar to that used for laser recrystallization (Figure 2.2.1). Using this technique, multilayer SOI recrystallization has been achieved to produce 3D IC's [2.77]. The second technique makes use of a periodic seeding structure. The scanning can be parallel, slanted or perpendicular to the seeding windows [2.72]. Back-side heating of the wafers is needed in order to reduce the thermal stress across the wafer during recrystallization. Both radiation heating from halogen lamps or a glowing filament, and heating from a second electron beam, directed at the back of the wafer, can be used [2.73, 2.78].

2.4. Zone-melting recrystallization

One of the main limitations of laser recrystallization is the small molten zone produced by the focussed beam, which results in a long processing time needed to recrystallize a whole wafer. Recrystallization of a polysilicon film on an insulator can also be carried out using incoherent light (visible or near IR) sources. In this case, a narrow (a few millimeters) but long molten zone can be created on the wafer. A molten zone length of the size of an entire wafer diameter can readily be obtained. As a result, full recrystallization of a wafer can be carried out in a single pass. Such a recrystallization technique is generally referred to as Zone-Melting Recrystallization (ZMR) because of the analogy between this technique and the float-zone refining process used to produce silicon ingots. The first method which successfully achieved recrystallization of large-area samples makes use of a heated graphite strip which is scanned across the sample to be recrystallized. The set-up is called "graphite strip heater" (Figure 2.4.1). A resistively heated graphite susceptor is used to raise the temperature of the entire sample up to within a few hundreds degrees below the melting temperature of silicon. Additional heating is locally produced at the surface of the wafer using a resistively heated graphite strip located a few millimeters above the sample and scanned across it [2.79, 2.80]. A typical sample is made of a silicon wafer on which a ≅ 1-2 μm-thick oxide is grown. A ≅ 0.5-1 μm-thick layer of LPCVD amorphous or polycrystalline silicon is then deposited, and the whole structure is capped with a 2 μm-thick layer of deposited SiO_2 covered by a thin Si_3N_4 layer. The capping layer helps minimizing mass transport and protects the molten silicon from contaminants (such as carbon from the strip heater). Recrystallization is carried out in a vacuum or an inert gas ambient in order to keep the graphite elements from burning.

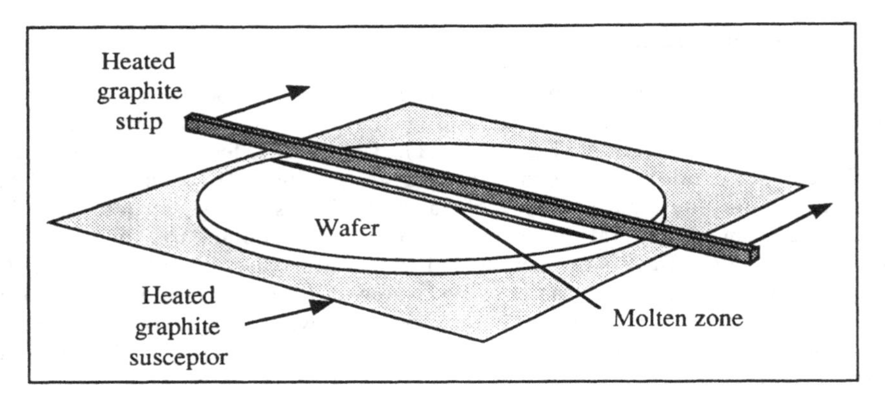

Figure 2.4.1: Zone-Melting Recrystallization of an SOI wafer using a graphite strip heater.

Both the graphite susceptor and the graphite strip can be replaced by lamps to achieve Zone Melting Recrystallization of SOI wafers. A lamp recrystallization system is composed of a bank of halogen lamps which is used to heat the wafer from the back to a high temperature (1100°C or above), and a top halogen or mercury lamp whose light is focussed on the sample by means of an elliptical reflector (Figure 2.4.2) [2.81-84]. A depolished quartz plate may be inserted between the lamp bank and the wafer in order to homogenize the energy deposition at the back of the wafer. As in the case of strip heater recrystallization, a narrow, wafer-long molten zone is created and scanned across the wafer with a speed on the order of 0.1-1 mm/sec.

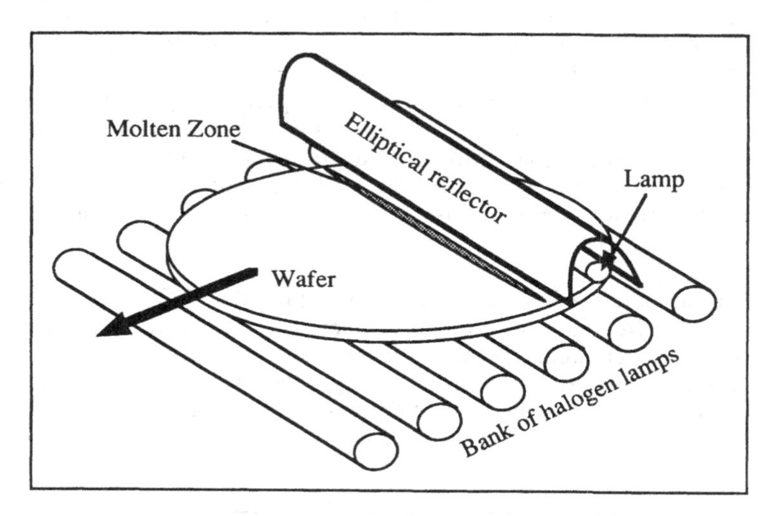

Figure 2.4.2: ZMR recrystallization of an SOI wafer using lamps.

The physics of zone-melting recrystallization is rather independent on the type of radiation source used (graphite strip heater or focussed lamp system), but it is quite different from that of laser recrystallization. Indeed, the dwell time (time during which a portion of silicon is exposed to the beam) is in the order of a millisecond or less in the case of cw laser processing, while the dwell time is in the order of seconds in the case of ZMR. The thermal gradients are also quite different: in a typical laser recrystallization experiment, the substrate is heated to 300-600°C, while ZMR uses preheating temperatures between 1100 and 1350°C.

2.4.1. Zone-melting recrystallization mechanisms

2.4.1.1. Melting front

Contrarily to what occurs in zone melting and recrystallization of bulk silicon, the crystalline "memory" of the silicon film is not completely erased upon melting, due to the thin-film configuration and to the presence of a stiff capping material which is capable to keep a morphological memory [2.85]. As a consequence, the final quality of the crystal shows some dependence on the melting front dynamics, and instabilities in the melting front can be responsible for defects in the recrystallized film. Upon zone-melting of thin SOI films (0.5 μm or less), the front solid/liquid interface does not advance continuously at the speed of scanning but progresses in a succession of bursts. This phenomenon is known as "explosive melting" [2.86]. This effect is attributed to the slightly different melting temperature of silicon grains having different crystal orientations and superheating in polycrystalline silicon. As a result, some solid silicon crystallites can be found in the molten silicon near the solid/liquid interface, and molten silicon droplets can be found ahead of the melting front. Sudden melting of superheated silicon grains causes the melting front to propagate in bursts (Figure 2.4.3).

Because bursts of melting are very rapid, the SiO_2 cap cannot flow and accommodate for the volume decrease of melting silicon, and bubbles can be created. These bubbles can give rise to voids in the recrystallized film. It is worth noting that this effect is inversely proportional to the silicon film thickness and is usually not observed during recrystallization of thick silicon films [2.85].

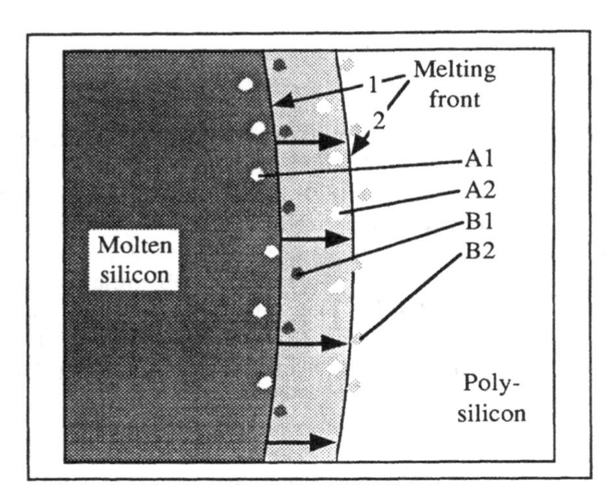

Figure 2.4.3: Evolution of melting front during ZMR processing (1 = at time 1 and 2 = at time 2 > time 1). A = solid crystallites in the molten zone, B = liquid silicon droplets ahead of the molten zone [2.85].

2.4.1.2 Solidification front

The solidification front is even more important than the melting front in determining the crystalline properties of the final SOI film. The recrystallized film has a (100) normal orientation, even if no seeding is used. Grain boundary-like defects repeated at 100 μm to several mm intervals are found in the recrystallized films. These defects can either be real grain boundaries, subgrain boundaries (*i.e.* low angle boundaries between adjacent (100) crystals or dislocation networks. Subgrain boundaries have a distinct wishbone pattern (Figure 2.4.4). In the best cases, the dislocation networks can be reduced to a few isolated dislocations.

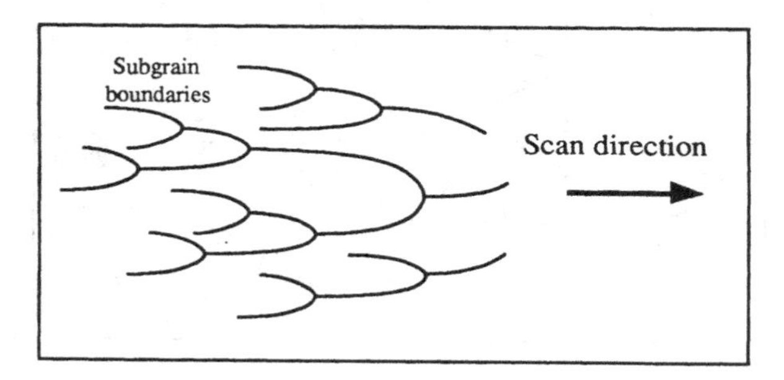

Figure 2.4.4: Pattern of subgrain boundaries in a ZMR film.

There is been quite a debate concerning the origin of these defects. Two mechanisms have been proposed, based either on a cellular growth (or cellular-dendritic growth) mechanism or on a faceted growth process. Cellular growth postulates the presence of a high concentration gradient of impurities such as nitrogen or carbon in the molten silicon [2.84, 2.87-90]. In that case, the material can be regarded as a dilute alloy which is locally constitutionally supercooled. The criterion for the existence of a constitutional supercooling is that $\frac{G}{R} < \frac{m\, C_o\, (1-k)}{k\, D}$ where R is the scan rate, G is the thermal gradient at the solidification interface, m is the liquidus slope of the silicon-impurity binary system [2.91], C_o is the impurity concentration in solid silicon, k is the solid-liquid impurity distribution coefficient ($k=7\times10^{-4}$ in the case of nitrogen), and D is the diffusion coefficient of the impurity in the liquid. If there is a region where the actual temperature is below the freezing temperature (constitutional supercooling), any perturbation of the interface will tend to grow as shown in Figure 2.4.5. Subgrain boundaries will appear at the trailing tips of the solidification interface, and their spacing will be comparable to the width of the supercooled region. This model can generally account for the spacing of the subgrain boundaries as a function of thermal gradients (but not as a function of scanning speed) and is best applicable when low thermal gradients are used (low thermal gradient regime). It also accounts for the rejection of impurities such as nitrogen towards the boundaries, which can be evidenced by techniques such as AES (Auger Electron Spectroscopy) mapping [2.84].

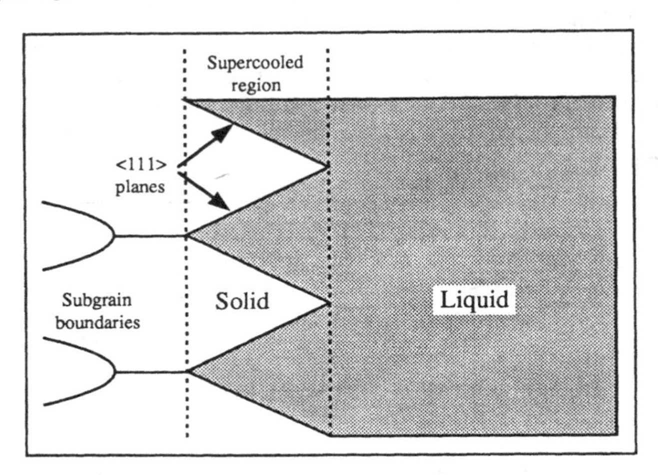

Figure 2.4.5: Cellular growth model: faceted protrusions span the undercooled region and form subgrain boundaries where they meet [2.89].

The cellular growth model, however is limited to those cases where there is a high concentration of impurities in the silicon film and when thermal gradients are extremely low, which is not always the case in real ZMR processing.

The second model is the faceted growth model [2.80, 2.93]. It is based on the three following assumptions: 1) the solid presents only <111> facets at the solid-liquid interface if the normal orientation of the film is (100) and scanning takes place in the <100> direction (<111> facets have the lowest interface energy and are the slowest growing planes), 2) the growth rate of each facet is limited by the nucleation rate of new atomic layers (which is dependent on the facet size - the smaller the facet, the larger the rate of addition of quenching atomic layers on it), and 3) a new re-entrant corner is formed as soon as a tip of the solid-liquid interface reaches the melting isotherm. Simulations of the dynamics of the ZMR solidification front have been carried out based on the three above assumptions [2.94]. These simulations show that the loci traced by the trailing tips of the liquid-solid interface form a branching pattern similar to the wishbone structure of the subgrain boundaries. The model accounts for the subboundary spacing as a function of the scan speed as well. The formation of subboundaries is illustrated in Figure 2.4.6 where the solidification front is depicted at three consecutive times.

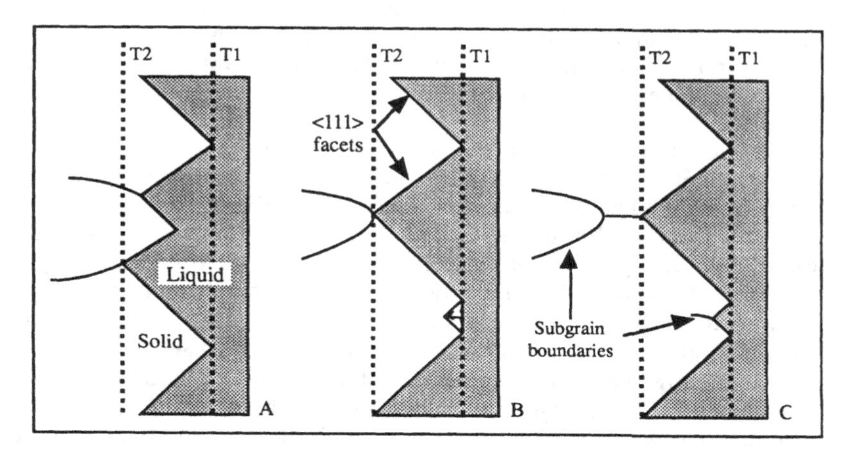

Figure 2.4.6: Faceted growth model: evolution of the solidification front at three consecutive times. Subboundaries are originated at the tips of the interface (A), the characteristic wishbone pattern is due to the coalescence of two tips (B), and a new <111> facets form at re-entrant corner of the melting isotherm, T_1 (C). T_2 is the solidification isotherm [2.80].

Boundaries originate at the trailing tips of the solidification interface. Their formation is related to a concentration of stress and a pile-up of defect-generating impurities dissolved into the silicon. The silicon film has a (100) normal orientation, with <100> direction parallel to the scan motion. Growth occurs by the addition of atoms at ledges or steps which sweep rapidly across the <100> facets. Therefore, it takes a longer time for adding a new layer of atoms on a long facet than on a short one. The distance a wedge-shaped pair of facets extends into the liquid region, which determines the spacing between the subgrain boundaries originating at the interior corners, is limited by two temperatures, T_1 and T_2, shown as dotted isothermal contours. T_2 is the temperature at which the generation of new ledges occurs at such a rate that the forward advance of the <111> facets exactly matches the imposed scanning velocity. At temperatures close to T_2 the speed with which ledges sweep across a <111> facet is much greater than the rate of forward advance of the solidification front. Therefore, the rate of forward advance is determined by the rate of generation of new ledges. (The velocity of the ledges decreases as a ledge encounters higher temperatures, and, eventually, the forward velocity of the ledge matches the scanning velocity when T_1 is reached.) Subboundaries tend to be parallel to the scan direction, but small perturbations cause them to move laterally. Quite often two subboundaries coalesce in a wishbone pattern and form a single subgrain boundary (Figure 2.4.6.C). When the distance between adjacent subboundaries becomes so large that the wedge formed by a pair of facets would extent beyond the T_1 isotherm, the interface becomes flattened (Figure 2.4.6.B). This flattened <100> interface is not stable, and two new <111> facets immediately develop, giving rise to a new subboundary (Figure 2.4.6.C).

This model accounts for the subboundary spacing as a function of thermal gradient (which influences the distance between T_1 and T_2), and scan speed. The higher the thermal gradient, the smaller the spacing between subboundaries [2.95]. The thermal gradient is controlled by the profile of energy deposited on the wafer by the incoherent light source. The misalignment between adjacent crystals has been found to decrease as a function of subboundary spacing, and is ruled by the following empirical law: $\Theta = 19.7\ S^{-0.9}$, where Θ is the angle of relative misalignment (in degrees), and S is the subboundary spacing (in micrometers) [2.95]. The faceted model is best applicable in cases where high thermal gradients are used (high-thermal gradient regime).

Actual ZMR processing involves both faceted and cellular growth (faceted growth appears to be prevalent near the melting isotherm, and cellular-dendritic growth is more important near the solidification isotherm) [2.85, 2.96]. Furthermore, more recent studies reveal that the establishment of an equilibrium between cooling of the high-reflectivity liquid silicon and heating of the lower-reflectivity solidified silicon can also explain the formation of a stable solidification interface which produces straight (unbranched) dislocation trails as only observable defect [2.96]. Indeed, when optimal conditions are used (i.e. when ultra-high purity silane is used for polysilicon deposition, when backside heating of the wafer is very uniform, when low thermal gradients are used ("low thermal gradient regime"), and when the scanning speed is thoroughly controlled, a stable solidification front can be obtained, subboundaries can be eliminated, and only parallel trails of isolated dislocations (*i.e.* without wishbone patterns) are observed in the SOI layer [2.96-98]. Such an "optimized" material exhibits average defect densities lower than 5×10^4 dislocations cm^{-2}, an intrinsic dopant level below 2×10^{15} cm^{-2}, junction leakages below 10^{-6} A/cm^2 and a minority carrier lifetime higher than 30 μs [2.97].

2.4.2. Role of the encapsulation

During the preparation of wafers for Zone-Melting Recrystallization, it is necessary to cap the polysilicon film with a layer which prevents the molten silicon from beading up. The tendency to bead up (or: delaminate, ball up) is a consequence of the poor wetting properties of molten silicon on SiO_2 [2.98-101]. Indeed, the wetting angle of molten silicon in a capillary sandwich of SiO_2 must be smaller than 90 degrees (Figure 2.4.7).

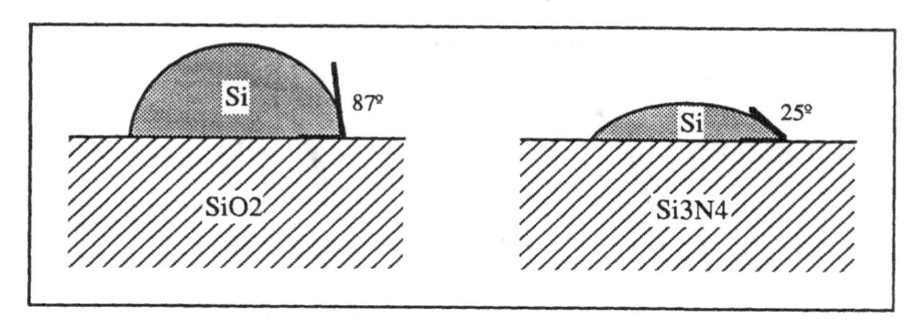

Figure 2.4.7: Wetting angle of molten silicon on SiO_2 and Si_3N_4 [2.100].

A wetting angle of 87 degrees is experimentally observed [2.102], which renders wetting possible, but extremely unstable. As a consequence, de-wetting and balling-up of the silicon occurs in most cases where a pure SiO_2 cap is used. The wetting angle of silicon on silicon nitride is much lower (25 degrees). If a thin ($\cong$ 5 nm) Si_3N_4 layer is deposited between the polysilicon film and the oxide cap, wetting is substantially improved, and reliable recrystallization can be achieved. Unfortunately, recrystallization using this much nitride does not produce (100)-oriented SOI films [2.100]. Reliable (100) recrystallization can be obtained by introducing less nitrogen at the silicon-cap interface (about one-third of a monolayer of nitrogen). This can be achieved by various methods: NH_3 annealing of the SiO_2 cap, deposition of SiN_x or Si_3N_4 on the cap, or plasma nitridation of the cap. These processes allow small amounts of nitrogen to diffuse towards the Si-SiO_2 interface during ZMR processing, which drastically improves wetting and, therefore, the quality of recrystallization.

2.4.3. Mass transport

The thickness uniformity of an SOI layer is of crucial importance, especially for thin-film applications. Indeed, some thin SOI MOSFET parameters such as threshold voltage and subthreshold slope are extremely sensitive to film thickness variations. SOI films produced by the ZMR technique exhibit two types of thickness non-uniformity. The first one is long-range mass transport caused by the sweeping of the molten zone from one side of the wafer to the other. The silicon film is slightly thinner where the recrystallization begins and slightly thicker where it ends [2.103]. The second one is a short-range waviness, perpendicular to the scan direction, and associated with the non-planar shape of the solidification front and the generation of subgrain boundaries (or trails of dislocations).

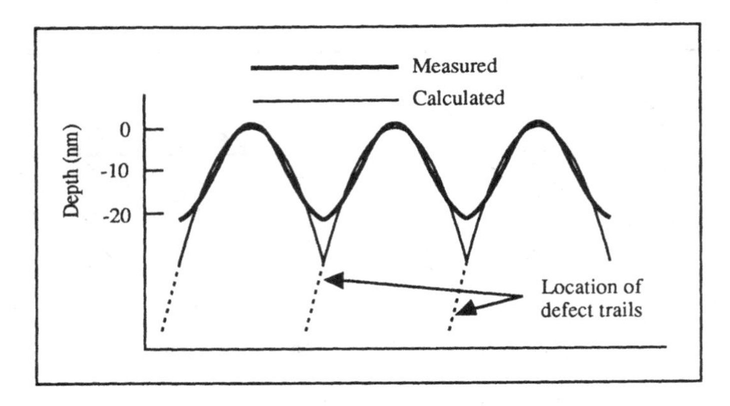

Figure 2.4.8: Amplitude of thickness variation measured perpendicular to the ZMR scan direction [2.103].

This short-range waviness is caused by the finite stiffness of the cap layer. For the "low-thermal gradient" regime (*i.e.* when non-branching, parallel and equally spaced dislocation trails are produced), the waviness is the result of the formation of a meniscus at the surface of the liquid silicon and of a surface tension equilibrium along the cap - solid silicon - liquid silicon triple line. The waviness of the recrystallized silicon film exhibits a periodical parabolic shape, the thickness reaching minima at the subgrain boundaries and maxima half way between the boundaries (Figure 2.4.8). A peak-to-peak thickness variation of ≅20 nm is measured on commercial ZMR wafers [2.103]. Because of film thickness variations and wetting problems during ZMR processing, recrystallization of thin (< 0.3 μm) films is unpractical.

2.4.4. Impurities in the ZMR film

Evidence of high contamination levels (such as SiC precipitates in subgrain boundaries) was reported in early ZMR material. Nowadays, the use of high-purity gases for polysilicon and SiO_2 deposition allow for the fabrication of electronic-grade ZMR films. The impurity found in highest concentration in the recrystallized silicon film is oxygen. The concentration of oxygen in the molten silicon during ZMR processing is 2.5×10^{18} atoms/cm^3 (oxygen solubility in liquid silicon). The segregation coefficient for oxygen at the solidification interface being close to 1, a similar concentration is found in the silicon film right after solidification.

During cooling down after ZMR processing, however, the oxygen dissolved in the silicon film segregates towards the Si-SiO_2 interfaces, which decreases the oxygen concentration in the silicon film. The effectiveness of this segregation is limited by the diffusion coefficient of oxygen in silicon to a fraction of a micron. As a result, the final peak oxygen concentration in thin ZMR silicon film is $\cong 10^{17}$ at./cm^3, while it is close to 2.5×10^{18} at./cm^3 in thicker films (Figure 2.4.9) [2.104].

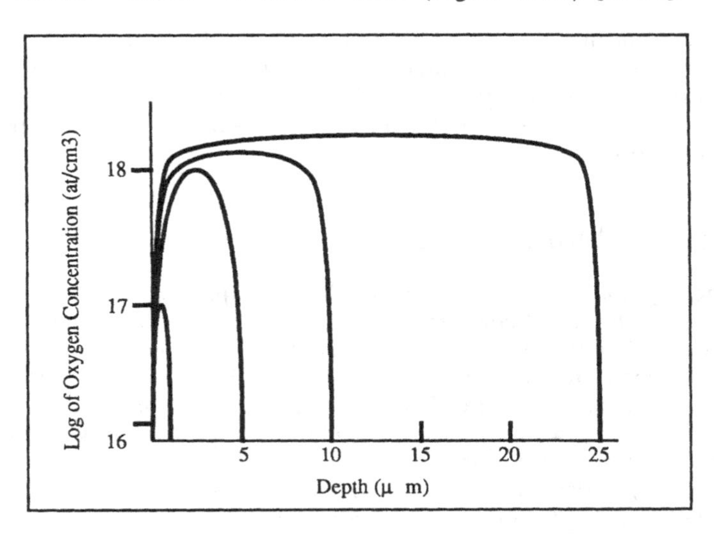

Figure 2.4.9: Oxygen concentration as a function of depth in 1, 5, 10, and 25 μm-thick ZMR silicon films.

The concentration of other impurities (carbon, transition metals, and shallow level impurities) is generally quite low in ZMR films, with typical values below 10^{16} cm^{-3}, as indicated by SIMS (Secondary Ion Mass Spectroscopy), AES (Auger Electron Spectroscopy), HIBS (Heavy Ion Back Scattering) and spreading resistance measurements [2.105-106].

Finally, one can mention that *thick* silicon films (10-100 μm) can be recrystallized over an insulator using the LEGO technique (Lateral Epitaxial Growth over Oxide)[2.107]. This technique makes use of a bank of stationary halogen lamps which heat the front side of a wafer. A thick polysilicon film is deposited on an oxidized silicon wafer where windows have been opened in the oxide for seeding purposes. As the temperature of the wafer is raised, the top polysilicon melts across the whole wafer. The temperature is then carefully ramped down, and crystal growth proceeds epitaxially from the seeding windows over the oxide until the entire silicon film is recrystallized. Recrystallization of thick silicon films can also be carried out on non-planar substrates using a scanning lamp apparatus [2.178]. This last method has been successfully applied to the realization of dielectrically isolated high-voltage MOSFETs [2.179].

2.5. Homoepitaxial techniques

Silicon-on-insulator can be produced by homoepitaxial growth of silicon on silicon, provided that the crystal growth can extend laterally on an insulator (SiO_2, typically). This can be achieved either using a classical epitaxy reactor or by lateral solid-phase crystallization of a deposited amorphous silicon layer.

2.5.1. Epitaxial lateral overgrowth

The Epitaxial Lateral Overgrowth technique (ELO) consists in the epitaxial growth of silicon from seeding windows over SiO_2 islands or devices capped with an insulator. It can be performed in an atmospheric or in a reduced-pressure epitaxial reactor [2.108-116]. The principle of ELO is illustrated in Figure 2.5.1. Typical sample preparation for ELO involves patterning windows in an oxide layer grown on a (100) silicon wafer. The edges of the windows are oriented along the <010> direction. After cleaning, the wafer is loaded into an epitaxial reactor and submitted to a high-temperature hydrogen bake to remove the native oxide from the seeding windows. Epitaxial growth is performed using *e.g.* an SiH_2Cl_2 + H_2 + HCl gas mixture. Unfortunately, nucleation of small silicon crystals with random orientation occurs on the oxide. These crystallites can be removed by an *in-situ* HCl etch step. Once the small nuclei are removed, a new epitaxial growth step is performed, followed by an etch step, and so on, until the oxide is covered by epitaxial silicon.

Epitaxy proceeds from the seeding windows both vertically and laterally, and the silicon crystal is limited by <100> and <101> facets (Figure 2.5.1.A). When two growth fronts, seeded from opposite sides of the oxide, join together, a continuous silicon-on-insulator film is formed, which contains a low-angle subgrain boundary where the two growth fronts meet. Because of the presence of <101> facets on the growing crystals, a groove is observed over the center of the SOI area (Figure 2.5.1.B). This groove, however, eventually disappears if additional epitaxial growth is performed (Figure 2.5.1.C) [2.108]. Three-dimensional stacked CMOS inverters have been realized by lateral overgrowth of silicon over MOS devices [2.115]. The major disadvantage of the ELO technique is the nearly 1:1 lateral-to-vertical growth ratio, which means that a 10 μm-thick film must be grown to cover 20 μm-wide oxide patterns (10 μm from each side). Furthermore, 10 additional micrometers must be grown in order to get a planar surface [2.108]. Thinner SOI films can, however, be obtained by polishing the wafers after the growth of a thick ELO film [2.216].

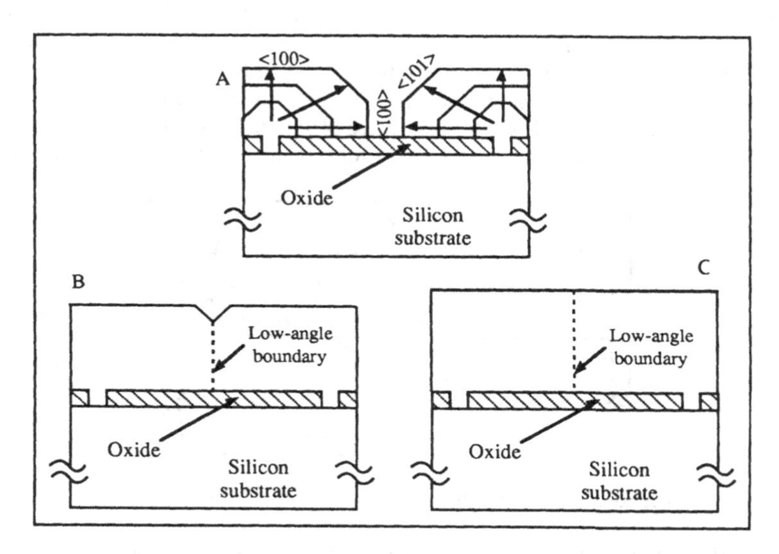

Figure 2.5.1: Epitaxial Lateral Overgrowth (ELO): growth from seeding windows (A), coalescence of adjacent crystals (B), self-planarization of the surface (C).

Recently, a variation of the ELO technique, called "tunnel epitaxy" or "confined lateral selective epitaxy "CLSEG", has been reported [2.117, 2.118]. In this technique, a "tunnel" of SiO_2 is created, which forces the epitaxial silicon to propagate laterally (Figure 2.5.2). With this method, a 7:1 lateral-to-vertical growth ratio has been obtained.

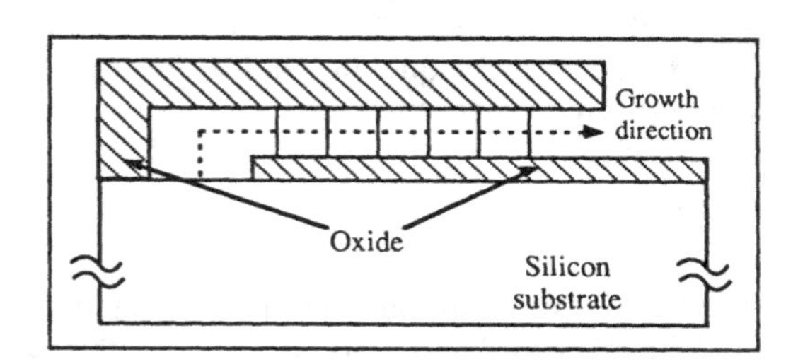

Figure 2.5.2: Principle of tunnel epitaxy [2.117].

2.5.2. Lateral solid-phase epitaxy

Lateral Solid-Phase Epitaxy (LSPE) is based on the lateral epitaxial growth of crystalline silicon through the controlled crystallization of amorphous silicon (α-Si) [2.119-125]. A seed is needed to provide the crystalline information necessary for the growth. The thin amorphous silicon film can either be deposited or obtained by amorphizing a polysilicon film by means of a silicon ion implantation step. LSPE is performed at relatively low temperature (575-600°C) in order to obtain regrowth while minimizing random nucleation in the amorphous silicon film.

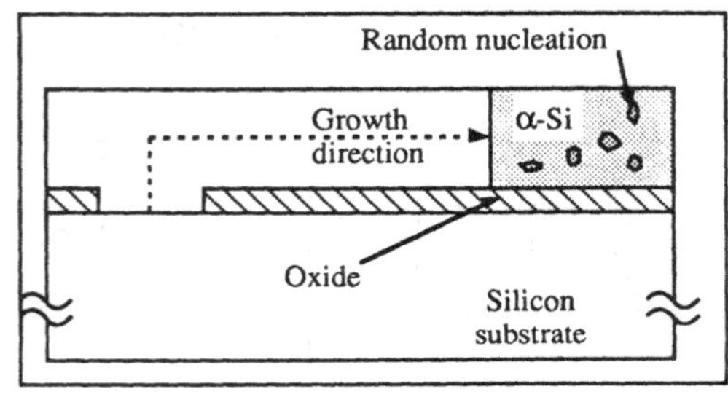

Figure 2.5.3: Principle of lateral solid-phase epitaxy.

The lateral epitaxy rate is in the order of 0.1 nm/s in undoped α-Si and 0.7 nm/s in heavily ($3x10^{20}cm^{-3}$) phosphorous-doped α-Si [2.122]. The distance over which lateral epitaxy can be performed over an oxide layer is in the order of 8 μm for undoped material, and 40 μm if a heavy phosphorus doping is used, further lateral extension of LSPE being limited by random nucleation in the α-Si film (Figure 2.5.3). An increase of the lateral extension of LSPE in undoped silicon can be obtained by alternating stripes of P-doped and undoped material, the stripes being perpendicular to the growth front (this can be performed through a mask step and ion implantation) [2.125]. This way, the growth in the doped material induces entrainment of the growth front in the undoped material. Research activity on LSPE is mainly carried out in Japan, where it is believed to be a technique of choice for the fabrication of 3D IC's using low-temperature processing.

2.6. FIPOS

The process of Full Isolation by Porous Oxidized Silicon (FIPOS) was invented in 1981 [2.126-2.127]. It relies on the conversion of a layer of silicon into porous silicon and on the subsequent oxidation of this porous layer. The oxidation rate of porous silicon being orders of magnitude higher than that of monolithic silicon, a full porous silicon buried layer can be oxidized while barely growing a thin oxide on silicon islands on top of it.

The original FIPOS process is described in Figure 2.6.1. It relies on the fact that p-type silicon can readily be converted into porous silicon by electrochemical dissolution of p-type silicon in HF (the sample is immersed into an HF solution and a potential drop is applied between the sample and a platinum electrode dipped into the electrolyte). The conversion rate of n-type silicon is much lower. Porous silicon formation proceeds as follows: at first, an Si_3N_4 film is patterned over a p-type silicon wafer, and boron is implanted to control the density of the porous silicon surface layer. The N^- material is formed by conversion of the P^- silicon into N^- silicon by proton (H^+ ion) implantation. The p-type silicon is then converted into porous silicon by anodization in a hydrogen fluoride solution. Optimal conversion yields a 56% porosity (porosity is controlled by the HF concentration in the electrolyte, the applied potential, and the current density at the sample surface during anodization). In this way, the volume of the buried oxide formed by oxidation of the porous layer is equal to that of the porous silicon, and stress in the films can be minimized.

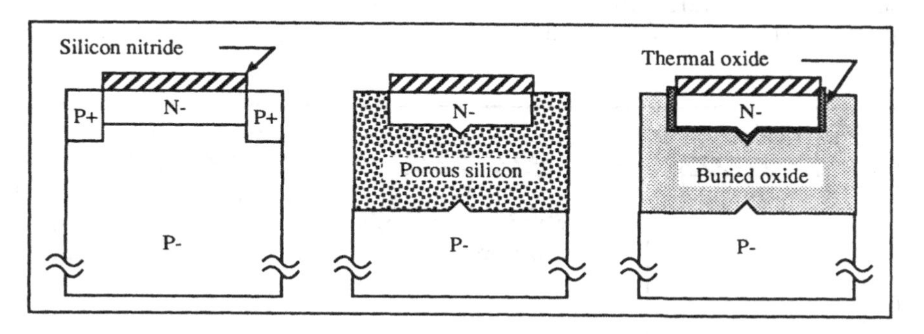

Figure 2.6.1: Original FIPOS process [2.126]. From left to right: formation of N^- islands and P^+ current paths, formation of porous silicon, and oxidation of the porous silicon.

Porous silicon contains an intricated network of pores. As a result, the area of silicon exposed to the outside world is extremely high, and porous silicon oxidizes very rapidly. This allows one to grow a thick buried oxide (oxidation depth is comparable to the width of the N^- silicon islands) while growing only a thin oxide at the edges of the N^- silicon islands in which the devices will be made. This last oxide being a thermally grown oxide, it provides the silicon islands with a high-quality bottom interface.

The original FIPOS technique produces high-quality SOI islands, and has been used to produce devices with good electrical characteristics [2.128, 2.129] but it has several limitations. Indeed, the formation of a thick oxidized porous silicon layer is needed to isolate even small islands. Such a thick oxide induces wafer warpage, especially if the islands are unevenly distributed across the wafer. A second limitation is the formation of a little cusp of unanodized silicon at the bottom-center of the silicon islands, where the two (left and right) anodization fronts meet during porous silicon formation (Figure 2.6.1). This limits the use of such a technique for thin-film SOI applications, where having a good thickness uniformity of the silicon islands is of crucial importance. Several variations of the process have been proposed, such as the formation of a buried P^+ layer (over a P^- substrate) located below N^- silicon islands, the whole structure being produced *e.g.* by epitaxy [2.130]. This technique solves the problem of having to produce a very thick buried layer to isolate wide islands, since it permits to obtain an island width-to-porous silicon layer thickness larger than 50.

A different approach for the fabrication of FIPOS structures is based on the preferential anodization of the N^+ layer of an $N^-/N^+/N^-$ structure (Figure 2.6.2) [2.131-133].

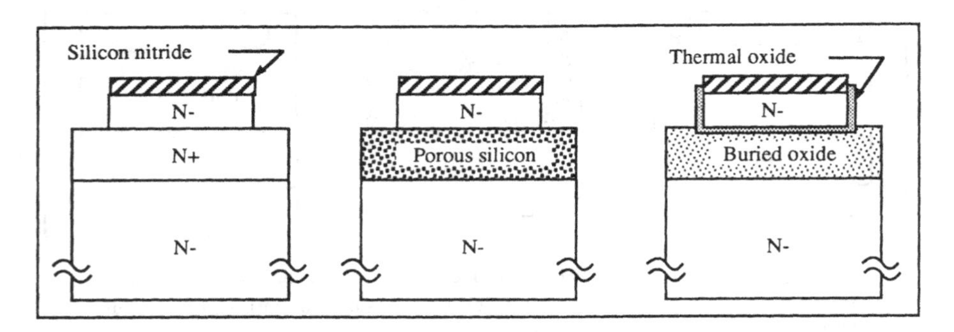

Figure 2.6.2: FIPOS formation ($N^-/N^+/N^-$ technique). From left to right: $N^-/N^+/N^-$ structure, formation of porous silicon, and oxidation of the porous silicon.

With this method, the thickness of both the silicon islands and the porous silicon layer are uniform and easily controlled by the N^+ doping profile (*e.g.* obtained by antimony implant on a lightly-doped, n-type wafer, and subsequent epitaxy of an N^- superficial layer). Another advantage of the $N^-/N^+/N^-$ approach is the automatic endpoint on the island isolation. As soon as all of the easily anodized N^+ layer is converted into porous silicon, the anodization current drops, and anodization stops due to a change in anodization potential threshold between the N^+ layer and the N^- silicon in the substrate and the islands. Such an automatic end of reaction control is not available in the $N^-/P^+/P^-$ approach where anodization of the P^- substrate occurs as soon as the P^+-layer has been converted into porous silicon, unless the anodization current is reduced to zero. After oxidation of the porous silicon layer [2.134], a dense buried oxide layer is obtained.

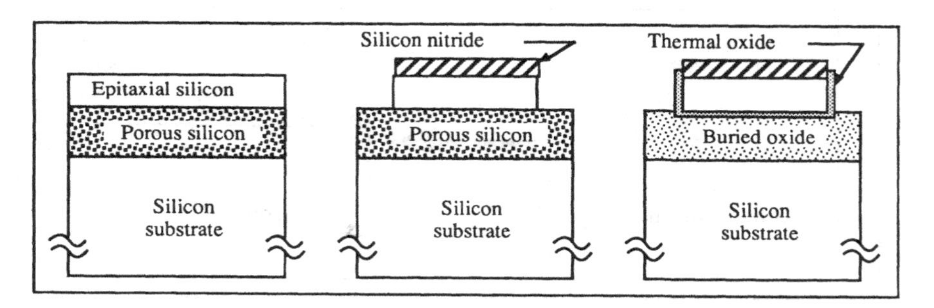

Figure 2.6.3: FIPOS formation by blanket anodization and epitaxy, trench etching and porous silicon oxidation (left to right).

It is worth noticing that porous silicon is a single crystal material, in spite of the fact that it contains many voids. A blanket porous silicon layer can, therefore , be created on a wafer, and "regular" single-crystal silicon can be grown onto it using MBE (molecular-beam epitaxy) [2.135] or low-temperature PECVD (plasma-enhanced chemical vapor deposition) techniques [2.136]. Trenches are then etched through the dense silicon overlayer in order to reach the buried porous silicon layer, the oxidation of the porous silicon can then be performed (Figure 2.6.3). This method has the advantage of easy blanket anodization, but it has the drawback of relying on delicate epitaxial procedures (MBE or PECVD). Large-scale integrated circuits (16k and 64k SRAMs) [2.137-138] as well as thin-film SOI devices [2.139] have been demonstrated using FIPOS technology. The technique, however, suffers from the fact that the anodization and the buried oxide formation are , in most cases, carried out after a mask step, which means that "FIPOS substrates" cannot be commercialized by a vendor.

2.7. SIMOX

The acronym SIMOX stands for "Separation by IMplanted OXygen". The principle of SIMOX material formation is very simple (Figure 2.7.1), and consists in the formation of a buried layer of SiO_2 by implantation of oxygen ions beneath the surface of a silicon wafer.

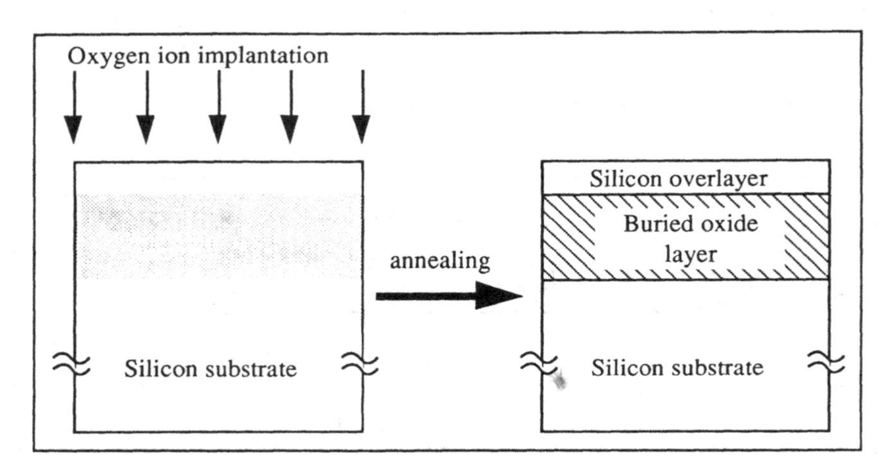

Figure 2.7.1: The principle of SIMOX: a heavy-dose oxygen implantation into silicon followed by an annealing step produces a buried layer of silicon dioxide below a thin, single-crystal silicon overlayer.

Processing conditions must be such that a single-crystal overlayer of silicon is maintained above the oxide. It is worth noticing that in conventional microelectronics, ion implantation is used to introduce atoms into silicon at the impurity level. Boron, for instance, is introduced in ppb (part per billion) quantities for threshold voltage adjust purposes, and the heaviest implants (S&D implants) introduce impurity concentrations of the order of one percent (for 99% of silicon atoms). In the case of SIMOX, ion implantation is used to synthesize a new material, namely SiO_2. This means that 2 atoms of oxygen have to be implanted for every silicon atom over a depth over which silicon dioxide has to be formed. In other words, the implanted dose required to form a buried oxide layer has to be ≅200-500 times higher than the heaviest doses commonly used for microelectronics processing. Indeed, O^+ doses of 1.8×10^{18} cm^{-2} are commonly used to produce SIMOX material, while doses of $\cong 10^{16}$ cm^{-2} are the usual upper boundary for dopant impurity implantations.

2.7.1. A brief history of SIMOX

Some of the important milestones in SIMOX technology are summarized in Table 2.7.1.

Year	Name/Company	Milestone	Ref.
1963	Watanabe & Tooi	Synthesis of SiO2 by oxygen implantation	2 .140
1978	Izumi et al. (NTT)	SIMOX acronym, First SIMOX circuits	2.141
1982	NTT	1k SRAM	2.142
1985	EATON - NTT	NV-200 high current oxygen implanter	
1985	LETI	High-temperature annealing (1300°C)	2.143
1986	AT&T	High-temperature annealing (1405°C)	2.144
1987	IBIS	SIMOX wafers commercial (3" to 6")	
1987	British Tel./Surrey	Masked implantation	2.145
1987	Hewlett-Packard	Thin-film (fully depleted) CMOS circuits	2.146
1987	Monsanto	Multiple implantation	2.147
1988	Hewlett-Packard	2 GHz thin-film CMOS circuits	2.148
1988	Harris	High-temperature CMOS 4k SRAM	2.149
1989	SPIRE	Low energy, low dose implants	2 .150
1989	TI / Harris	64k SRAM	2.151 2.152
1989	LETI	Thin-film 16k SRAM	2.153
1989	AT&T	6.2 GHz thin-film CMOS circuits	2.154
1989	NTT	21 ps CMOS ring oscillator	2.155
1990	TI	Commercial 64k SRAM	

Table 2.7.1: Some milestones of SIMOX technology.

The first SiO_2 layer synthesized by ion implantation of oxygen into silicon was reported in 1963, and in 1977, the first buried oxide layer was produced by the same technique. A year later, the acronym of SIMOX was proposed for this technique, and 19-stage ring oscillators were fabricated in that material. At that time, most of the SOI research worldwide was concentrating on recrystallization techniques, and the newcoming SIMOX technique was considered quite exotic, and researchers were skeptical about its possible use in microelectronics. Indeed, early SIMOX material was of poor quality. The defect density (dislocations, oxide precipitates and polycrystalline silicon inclusions) was so high that epitaxial silicon had to be grown on top of the silicon overlayer in order to obtain a silicon layer good enough for device fabrication. In addition, the implantation of a mere 2x2 cm^2 area on a silicon wafer could take on the order of 24 hours, using a conventional ion implanter. Nonetheless, a 1k SRAM was fabricated by NTT in 1982, and the SOI community started to look at this material with less skeptical eyes.

In 1985, the first NV200 high-current oxygen implanter was produced by Eaton according to NTT's specifications. This machine was designed to deliver an O^+ ion beam of up to 100 mA at energies of up to 200 keV. The availability of such a machine was the key to the successful development of SIMOX technology.

Before 1985, SIMOX material was usually annealed at 1150°C (which is the highest temperature available in conventional furnaces equipped with quartz tubes). In 1985, it was demonstrated that annealing at 1300°C in a furnace equipped with either a polysilicon or a silicon carbide tube, and even annealing at a few degrees below the melting point of silicon (1405°C) in a lamp annealing system could dramatically improve the quality of the material and produce atomically sharp silicon-oxide interfaces.

In 1987, it was discovered that the use of multiple implantation/annealing steps could be used in place of a single, high-dose implantation could dramatically decrease the dislocation density in the silicon overlayer (less than 10^3 dislocations/cm^2). Improvements of the NV200 implanter yielded steadily better material, with less metal and carbon contamination such that, for instance, 64k SRAMs can nowadays be produced with excellent yield in SIMOX material with standby current consumption as low as 10 nA [2.152].

Recent developments include low-dose, low-energy implantation for production of thin buried oxides and silicon overlayers with low defect density [2.150]. The research for producing SIMOX material using low-dose, low-energy implantation is also driven by economical reasons, since the production costs of SIMOX material are proportional to the dose and the energy used for the implantation.

2.7.2. Oxygen implantation

The shape and quality of SIMOX layers obtained by oxygen implantation depend on both the oxygen dose and the temperature of the wafer during the implantation process.

2.7.2.1. Oxygen dose

When a heavy dose of oxygen ions is implanted into silicon, several effects occur. These effects depend on the implanted dose. Stoechiometric SiO_2 contains 4.4×10^{22} oxygen atoms/cm^3. Therefore, the implantation of 4.4×10^{17} atoms/cm^2 should be sufficient to produce a 100 nm-thick buried oxide layer. Unfortunately, due to the statistical nature of ion implantation, the oxygen profile in silicon does not have a box shape, but rather a skewed gaussian profile, and the implanted atoms spread over more than 100 nm, such that SiO_2 stoichiometry is not reached (Figure 2.7.2).

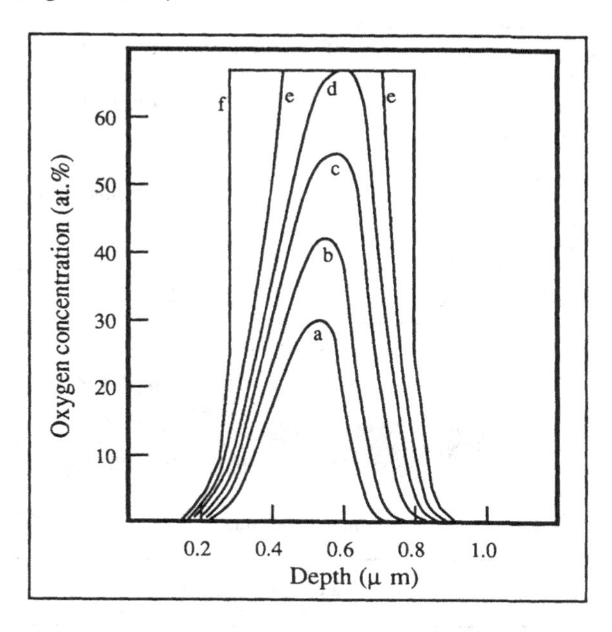

Figure 2.7.2: Evolution of the oxygen concentration profile with the implanted dose: a) $4x10^{17}$ cm^{-2}, b) $6x10^{17}$ cm^{-2}, c) 10^{18} cm^{-2}, d) $1.2x10^{18}$ cm^{-2}, e) $1.8x10^{18}$ cm^{-2}, and f) $2.4x10^{18}$ cm^{-2}. Energy is 200 keV [2.157].

If the wafer is annealed after such a "low-dose" implant, oxide precipitates will be formed at a depth equal to the depth of maximum oxygen concentration, but no continuous layer of SiO_2 will be formed. Experiments show that a dose of $1.4x10^{18}$ cm^{-3} has to be implanted in order to create a continuous buried oxide layer. The standard dose which is used by most people is $1.8x10^{18}$ cm^{-3}, which produces a 400 nm-thick buried oxide layer upon annealing. Figure 2.7.2 describes the evolution of the profile of oxygen atoms implanted into silicon with an energy of 200 keV. At low doses, a skewed gaussian profile is obtained. When the dose reaches $1.2...1.4x10^{18}$ cm^{-2}, stoichiometric SiO_2 is formed (66 at.% of oxygen for 33 at.% of silicon), and further implantation does not increase the peak oxygen concentration, but it rather broadens the overall profile (*i.e.* the buried oxide layer becomes thicker). This is possible because of the diffusivity of oxygen in SiO_2 (10^{17} $cm^2.s^{-1}$ at 500°C [2.158]) is high enough for the oxygen to readily diffuse to the Si-SiO_2 interfaces and oxidize the silicon. The dose at which the buried oxide starts to form (≅ $1.4x10^{18}$ cm^{-2}) is called "critical dose". Doses below or above that level are called "subcritical" or "supercritical", respectively. It is important to notice that a very large amount of oxygen is introduced into the silicon. Since the volume of synthesized SiO_2 is larger than the volume of silicon consumed to form the buried SiO_2 layer, swelling of the wafer is observed. In addition, significant sputtering of the top silicon layer is observed when the implanted dose exceeds 10^{18} cm^{-2} (Figure 2.7.3).

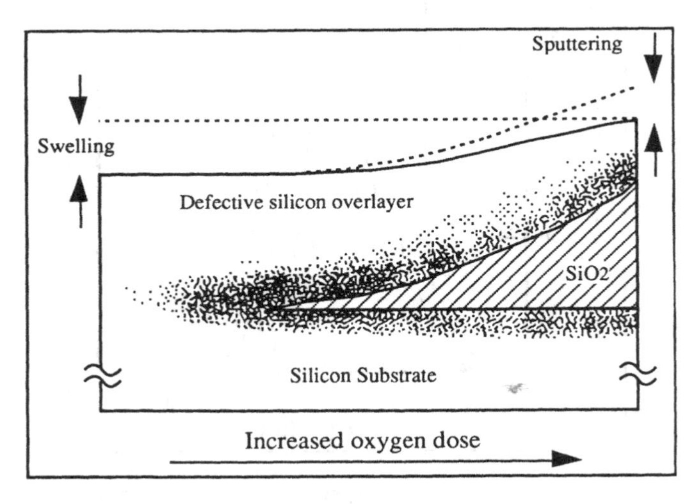

Figure 2.7.3: Evolution of the buried oxide layer formation as a function of implanted dose. From left to right: creation of a damaged layer into the silicon, amorphization of the silicon, creation of a buried oxide layer. Swelling and sputtering effects are also represented [2.159].

The sputtering rate ranges between 0.1 to 0.2 sputtered Si atom per implanted oxygen atom. Sputtering can be minimized by growing an SiO_2 layer on the silicon wafer prior to the implantation step. Swelling makes the silicon overlayer appear thicker than it really is, while sputtering thins it down.

2.7.2.2. Implant temperature

The temperature at which the implantation is performed is also an important parameter which influences the quality of the silicon overlayer. Indeed, the oxygen implantation step does amorphize the silicon which is located above the projected range. If the temperature of the silicon wafer during implantation is too low, the silicon overlayer gets completely amorphized, and it forms polycrystalline silicon upon further annealing, which is totally undesirable. When the implantation is carried out at higher temperatures (above 500°C)-(Figure 2.7.4), the amorphization damage anneals out during the implantation process ("self annealing"), and the single-crystallinity of the top silicon layer is maintained. The silicon overlayer, however, is highly defective.

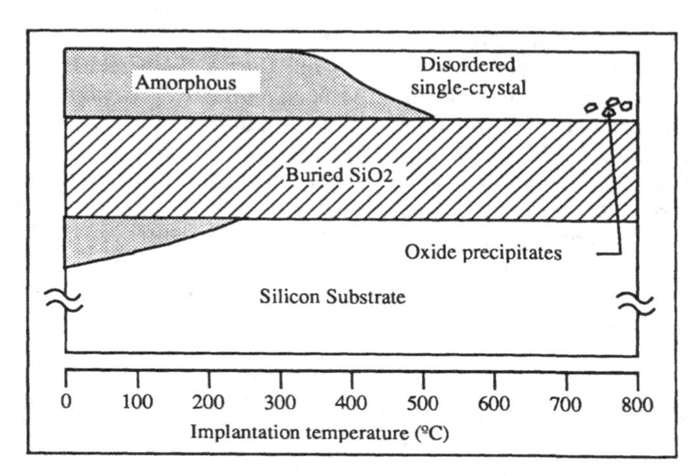

Figure 2.7.4: Evolution of the crystallinity of the SIMOX structure as a function of implantation temperature (dose = 1.8×10^{18} cm^{-2}, energy = 200 keV) [2.158].

At higher implant temperatures (700-800°C), SiO_2 precipitates form in the silicon overlayer, mostly near its bottom interface, although large precipitates have been observed near the surface region of the silicon film. This effect seems to set an upper limit to the implantation temperature to approximately 700°C. Implantation temperatures most commonly used range between 600°C and 650°C. Wafer heating during implantation is automatically achieved by the energy deposited into the wafer if a high-current implanter is used (beam current of 30-50 mA). If a conventional, low-current implanter is employed, wafer heating can be provided by a bank of halogen lamps or a heated chuck.

2.7.3. Annealing parameters

After the oxygen ion implantation has been performed, a thermal annealing step is necessary to form a device-worthy SIMOX structure. Generally speaking, the higher the annealing temperature, the better the quality of the material. Until 1985, temperatures of 1150°C were used. The obtained material was of rather poor quality (it had $\cong 10^9$ dislocations per cm^2). Later, furnace annealings at temperatures of 1250°C and 1300°C were achieved, which greatly improved the quality of the silicon overlayer [2.160]. Annealings were even carried out in lamp annealing systems at temperatures up to 1405°C [2.144]. 6-hour annealings at 1300-1350°C in furnaces equipped with either polysilicon or silicon carbide tubes are nowadays considered as a standard. Annealing is usually carried out in a nitrogen with 2% oxygen gas ambient. The oxygen allows for the growth of some oxide on the superficial silicon layer which protect the silicon from the pitting phenomenon which occurs when silicon is annealed at high temperature in pure nitrogen. A pure nitrogen ambient can be used as well, provided that a protecting CVD oxide layer has been deposited on the wafer prior to annealing. The annealing can be performed in an argon gas ambient as well. Better material qualities seem to be obtained when the annealing is carried out in argon (rather than in nitrogen). The kinetics of the annealing of the SIMOX structure is quite complicated and has been investigated by several research groups. We shall here describe only the results which are of immediate practical interest.

The evolution of the SIMOX structure upon different annealing steps is presented in Figure 2.7.5. We will take the example of the implantation of 1.5×10^{18} oxygen ions/cm^2 at 200 keV into silicon [2.160]. Three regions can be distinguished in the as-implanted sample: a 420 nm-thick, highly disordered but nonetheless single-crystal top silicon layer containing SiO_2 precipitates, the size of which increases from the top surface to the Si-SiO_2 interface, a 180 nm-thick amorphous buried oxide, and a heavily damaged silicon layer extending ...450 nm... into the substrate (Figure 2.7.5.A).

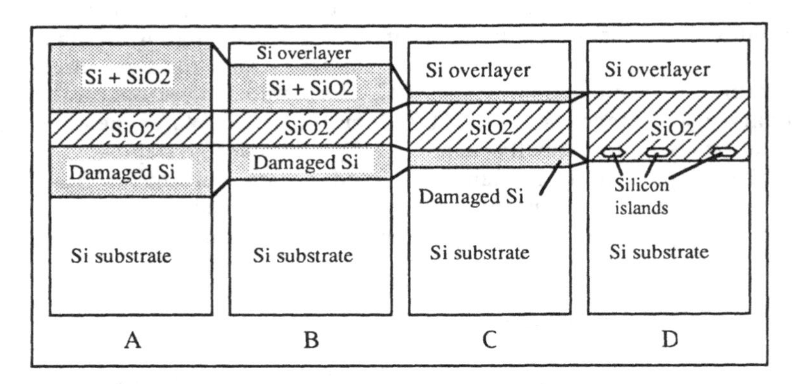

Figure 2.7.5: Evolution of the structure of the SIMOX structure as a function of post-implantation annealing temperature (implant dose=1.5 10^{18} cm^{-2}, energy=200 keV). A: As-implanted; B: 2-hour annealing at 1150°C; C: 6-hour annealing at 1185°C, and; D: 6-hour annealing at 1300°C [after 2.160].

After annealing at 1150°C, the first 80 nm of the top of the silicon overlayer are denuded from SiO_2 precipitates (Figure 2.7.5.B). The lower part of the silicon layer is highly defective and contains a large density of oxide precipitates with dislocations pinned onto them. The mean diameter of these oxide precipitates is approximately 25 nm. The buried oxide - silicon substrate interface region displays a lamellar structure (layered mixtures of Si and SiO_2). After annealing at 1185°C, the top 200 nm of the silicon layer are free from oxide precipitates (Figure 2.7.5.C), but large SiO_2 precipitates are found closer to the Si layer-buried oxide interface. These precipitates are fewer but larger than in the case of the 1150°C annealing. The lower SiO_2-Si substrate interface still shows a lamellar structure. Upon annealing at 1300°C, the whole silicon overlayer is denuded from oxide precipitates, and the top silicon-buried oxide interface is atomically sharp (Figure 2.7.5.D). Small silicon islands (remnants of the lamellar structure) are found within the buried oxide at a distance of 25 nm from the bottom interface. These islands have the same crystal orientation as the substrate, and they are 30 nm thick and 30 to 200 nm long. The evolution of the SIMOX structure can be explained by thermodynamical considerations. During annealing, both dissolution of the oxide precipitates and precipitation of the dissolved oxide take place in the oxygen-rich silicon layers. In order to minimize the total surface energy of the SiO_2 precipitates, small precipitates dissolve into silicon, and large precipitates grow from the dissolved oxygen. At any given temperature (and for a given concentration of oxygen in the silicon), there exists a critical precipitate radius, r_c, below which the precipitate will disappear, and above which it will be stable. It can be shown that $r_c = -\frac{2\sigma}{\Delta H_v}\frac{T_E}{T_E - T}$, where T is the temperature, T_E is the temperature of equilibrium between the two phases, ΔH_v is the volume enthalpy of formation of the SiO_2 phase, and σ is the surface energy of the precipitates. The critical radius depends both on the temperature and on T_E-T which is an image of the interstitial oxygen supersaturation. It also increases with temperature and becomes essentially infinite at very high temperatures (*i.e.* when $T \rightarrow T_E$), such that the only stable precipitate is the buried oxide layer, which has an infinite radius of curvature. This phenomenon of growth of the large precipitates at the expense of smaller ones is known as the "Ostwald ripening" [2.158, 2.160]. At relatively low annealing temperatures, the dissolution of small precipitates near the surface will contribute to the formation of a denuded zone at the top of the silicon overlayer, while larger precipitates will remain stable deeper into the silicon film (Figure 2.7.5.B). The complete dissolution of all precipitates upon high temperature annealing ($\geq$ 1300°C) and the diffusion of the dissolved oxygen towards the buried layer explain the thickness increase of the latter when high-temperature annealing is performed.

2.7.4. Multiple implants

The SIMOX implantation process introduces damages and stress into the silicon overlayer, such that massive amounts of crystalline defects (dislocations) are produced. Unlike oxide precipitates, dislocations are not eliminated by high-temperature annealing. It can be shown that the creation of some of these microdefects can be related to the introduction of impurities such as carbon into the silicon film during implantation [2.161], and the improvements of the cleanliness and the design of oxygen implanters has contributed to a steady reduction of the dislocation densities [2.162]. In 1987, it was observed that the creation of defects can be drastically minimized if the implanted dose remains below a threshold value of approximately 4×10^{17} ions cm^{-2} (at 150 keV). This threshold is unfortunately subcritical, such that no continuous buried oxide layer can be formed upon annealing. Oxide precipitates are formed, however, in the vicinity of the projected range of the oxygen ions (*i.e.* at a depth corresponding to the peaks of the gaussian profiles of Figure 2.7.2). If the implant and annealing processes are repeated a second and a third time (such that a total dose of *e.g.* 1.2×10^{18} atoms/cm^2 is implanted), high-quality SIMOX material can be produced, without generating significant amounts of dislocations in the silicon overlayer (dislocation densities as low as 10^3 cm^{-2} have been demonstrated). In addition to the low defect density, atomically sharp Si-SiO_2 interfaces are produced, and the bottom of the buried oxide is free of silicon islands [2.147].

2.7.5. Low-energy implantation

Recent SIMOX material research is oriented towards the fabrication of thin silicon films on thin buried oxides using low-energy, low-dose implantation. The drive for such a development is three-fold: firstly, the total-dose hardness of thin buried oxides is expected to be better than that of thicker ones. Secondly, direct fabrication of thin silicon films is attractive for thin-film device applications (instead of thinning the film down after thicker material fabrication). Finally, the production cost of SIMOX wafer is proportional to both the beam energy and to the implanted dose. It has been shown that the implantation of subcritical doses at energies ranging between 30 to 80 keV and followed by a high-temperature annealing step can produce thin and continuous buried oxide layers as well as thin silicon overlayers with low defect densities [1.150]. For instance, the implantation of 1.5×10^{17} O^+ ions cm^{-2} at 30 keV produces a 57 nm-thick silicon layer on top of a 47 nm-thick buried oxide layer. A potential additional benefit from this technique is the reduction of contamination of the wafers, since the introduction of impurities (carbon, heavy metals) is proportional to the implanted oxygen dose.

2.7.6. Material quality

The evolution of the dislocation density with time in SIMOX is reported in Figure 2.7.6 (adapted from [2.163]). The improvements are attributed to the better control of the implant conditions (cleanliness, stability of the beam, optimization of scanning techniques, and uniformity of wafer heating), and, more recently, to the introduction of multiple implant and anneal procedures.

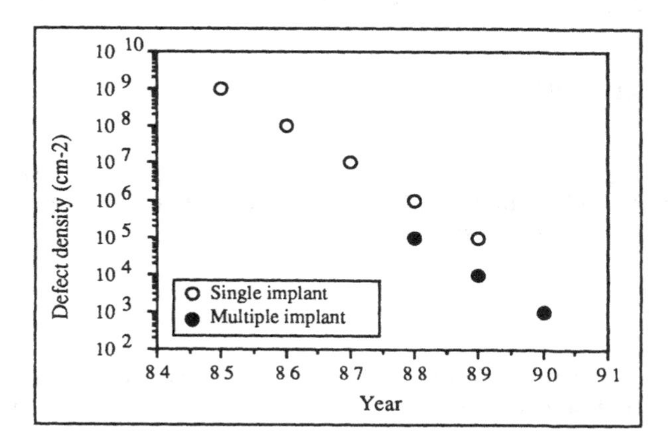

Figure 2.7.6: Evolution of the dislocation densities in SIMOX with time [2.163].

Early SIMOX material contained high concentrations of light metals such as aluminum and of heavy metals, such as iron, chromium and copper. These where sputtered from the implanter walls or the rotating drum on which the wafers were placed onto the wafers and thermally driven into the silicon by the wafer heating and the subsequent thermal annealing step. Such metallic impurities increase the leakage currents of junctions made in SIMOX material and degrade the radiation hardness of SIMOX MOSFETs. Great efforts have been made to reduce the metal contamination sources, and silicon or silicon dioxide shields have been placed in the implanters such that the beam never "sees" any metallic part [2.162]. The design of high-temperature furnaces used to anneal the SIMOX material has also been optimized in order to minimize the diffusion of metallic contaminants from the heating elements into the furnace tubes. Nowadays, metallic impurity concentrations below 10^{17} cm^{-3} are measured in SIMOX material [2.164], and it is estimated that concentrations below 10^{15} cm^{-3} will be within reach in a near future. Some of the characteristics of modern SIMOX material are listed in Table 2.7.2, as well as these which appear to be within reach in a near future.

	Present	Future	Units
Silicon film thickness uniformity	+/- 10	+/- 2.5	nm
Buried oxide thickness	300-500	50-500	nm
Buried oxide thickness uniformity	+/- 5	+/- 1	%
Defect density	1e4 - 1e5	1e2 - 1e3	cm-2
Particulates (> 1 micrometer)	0.75	0.016	cm-2
Carbon contamination	< 5e17	< 1e17	cm-3
Metal contamination	< 1e17	< 1e15	cm-3

Table 2.7.2: Characteristics of modern SIMOX material [2.162].

2.8. SIMNI and SIMON

In analogy to the SIMOX acronym, SIMNI and SIMON mean "Separation by IMplanted NItrogen" and "Separation by IMplanted Oxygen and Nitrogen", respectively.

2.8.1. SIMNI

Just as buried oxide can be synthesized by oxygen ion implantation and annealing, a buried silicon nitride layer (Si_3N_4) can be obtained by implanting nitrogen ions into silicon and annealing the structure. The critical dose for forming a buried nitride layer is 1.1×10^{18} cm^{-2} at 200 keV [2.158], but lower doses can be employed if the implant energy is lower. The obtained buried nitride and silicon overlayer thicknesses are 190 nm and 215 nm, respectively, in the case of a 7.5×10^{17} N^+ ions cm^{-2} implantation at 160 keV followed by a 1200°C anneal [2.165]. As a result of the lower dose needed to form a buried layer than in the case of SIMOX (early material, single implant), lower densities of dislocations are observed (< 10^7 cm-2). As in the case of SIMOX, the profiles of nitrogen atoms implanted into silicon can be modelled [2.166]. The most significant difference between oxygen and nitrogen profiles is that the peak of the nitrogen distribution does not saturate once stoichiometry is attained. This is due to the low diffusion coefficient of nitrogen in Si_3N_4 (10^{-28} $cm^2 s^{-1}$ at a temperature of 500°C against 10^{-17} $cm^2 s^{-1}$ for O_2 in SiO_2 [2.158]). The result of this low diffusion coefficient is that both nitrogen bonded to silicon and "unbonded" nitrogen are found in the buried layer if supercritical doses are implanted. During implantations, nuclei of crystalline silicon nitride are formed. Upon annealing (*e.g.* at 1200°C), nitride crystallites grow radially outwards from the nucleation sites, in a dendritic manner, by gettering nitrogen from the surrounding silicon, and polycrystalline α-

Si_3N_4 is formed. As in the case of SIMOX, low-dose, low-energy implants (60 keV, 7×10^{17} cm^{-2}) can be used to produce thinner buried dielectric layers [2.167].

The formation of buried nitride was presenting advantages over early SIMOX material: use of lower implant doses, non-corrosion of hot-filament implanter sources, and lower defect densities in the silicon overlayer. Unfortunately, nitride buried layers are polycrystalline (buried oxide layers are amorphous), and the grain boundaries between the Si_3N_4 crystallites are at the origin of leakage currents between devices made in the silicon overlayer and the underlying silicon substrate. Furthermore, Si-Si_3N_4 interfaces are known to be of much worse quality (higher density of surface states) than Si-SiO_2 interfaces. MOS device fabrication has, however, been demonstrated in SIMNI material [2.168].

2.8.2. SIMON

The implantation of both oxygen and nitrogen ions into silicon has been achieved by several research groups. These were attempting to combine the advantages of both SIMNI (formation of a buried layer by implantation of a relatively low dose and low defect generation) with those of SIMOX (formation of an amorphous rather than polycrystalline buried layer and good Si-dielectric interfaces) [2.167, 2.169]. Buried oxynitride layers may also present better radiation hardness performances than pure SIMOX material.

Buried oxynitride layers have been formed by implantation of different doses of both nitrogen and oxygen into silicon. Different implant schemes have been proposed (oxygen can be implanted before nitrogen, or vice-versa). The kinetics of synthesis of oxynitrides by ion implantation is more complicated than that of pure SIMOX or pure SIMNI materials. In some instances, gas (nitrogen) bubbles can be formed within the buried layer [2.169]. It is, however, possible to synthesize buried oxynitride layers which are stable and remain amorphous after annealing at 1200°C [2.167]. The resistivity of the buried oxynitride layers can reach up to 10^{15} Ω.cm, which is comparable to that of oxynitride layers formed by other means.

2.9. Wafer bonding and etch-back

The principle of the bonding and etch-back technique is very simple: two oxidized wafers are "glued" -or- "bonded" together. One of the wafers is subsequently polished or etched down to a thickness suitable for SOI applications. The other wafer serves as a mechanical substrate, and is called "handle wafer" (Figure 2.9.1).

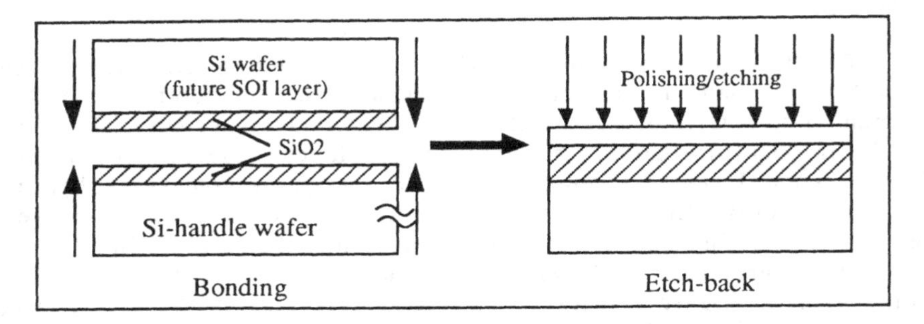

Figure 2.9.1: Bonding of two oxidized silicon wafers (left), and polishing/etching back of one of the wafers.

2.9.1. Mechanisms of bonding

When two flat, hydrophilic surfaces such as oxidized silicon wafers are placed against one another, bonding naturally occurs, even at room temperature. The contacting forces are believed to be caused by the attraction of hydroxyl groups $(OH)^-$ adsorbed on the two surfaces. This attraction can be significant enough to cause spontaneous formation of hydrogen bonds across the gap between the two wafers [1.170]. This attraction propagates from a first site of contacting across the whole wafer in the form of a "contacting wave" with a speed of several cm/s (Figure 2.9.2) [1.171].

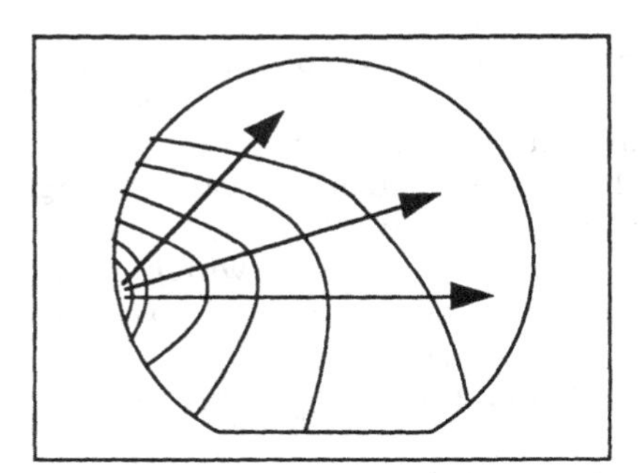

Figure 2.9.2: Propagation of a contacting wave.

After the wafers have been bonded at room temperature, it is customary to anneal the structure to strengthen the bonding. The bond strength increases with annealing temperature, and three phases of bond strengthening occurring at different temperatures can be distinguished. The first phase (bonding between hydroxyl groups), starts to be replaced by Si-O-Si bonds around 300°C. At higher temperatures (1100°C and above), the viscous flow of the oxide allows for complete "welding" of the wafers. The quality of bonding depends on the roughness of surfaces in contact. For example, it has been shown that much more uniform bonding is obtained if the roughness of the bonded wafers is less than 0.5 nm than if it is larger than that value [2.172].

When the wafers are annealed at 300°C and above, the following reaction takes place : Si-OH + OH-Si $\rightarrow$ Si-O-Si + H_2O [2.172]. As a result of water vapor formation during this reaction, voids appear between the bonded wafers. These voids are called "intrinsic voids" and they disappear at higher temperatures, as the water reacts and oxidizes the silicon (as in a thermal wet oxidation process). The hydrogen produced by the latter reaction diffuses easily out of the wafers.

Another type of voids, called "extrinsic voids" are caused by the presence of particulates between the wafers (Figure 2.9.3). Voids with a diameter of several millimeters are readily created by particles 1 μm or less in size. Because of the usually organic nature of the particulates, the extrinsic voids cannot be eliminated by high-temperature annealing. The only way of obtaining void-free bonding is, therefore, to clean the wafers with ultra-pure chemicals and water and to carry out all bonding operations in an ultra-clean environment. The presence of voids is easily revealed by infrared or ultrasonic imaging, as well as by X-ray tomography [1.172]. The presence of voids is, of course, undesirable, since they can lead to delamination of devices located in imperfectly bonded areas during device processing.

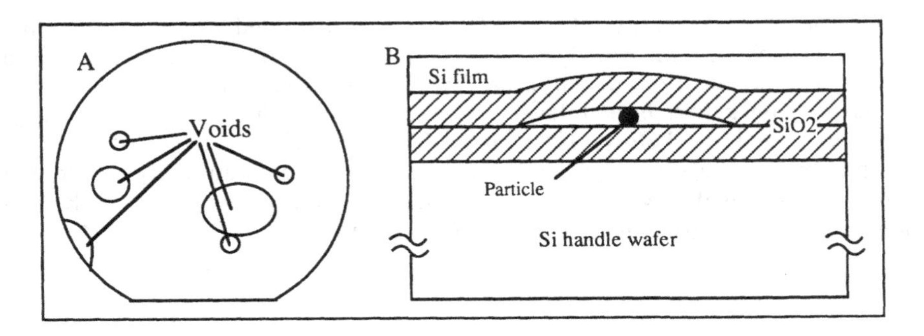

Figure 2.9.3: Extrinsic voids, as seen by infrared, ultrasonic or X-ray imaging of bonded wafers (A), and cross-section of an extrinsic void caused by a particle (B).

2.9.2. Etch-back techniques

After bonding of the wafers has been carried out, the top wafer has to be thinned down from a thickness of ...600 μm... down to a few micrometers or less in order to be useful for SOI device applications. Two basic thinning approaches can be used: grinding followed by chemico-mechanical polishing, and grinding followed by selective etch-back. The grinding operation is a rather crude but rapid step which is used to remove all but the last several micrometers of the (top) bonded wafer. The thinning method using chemico-mechanical polishing is cheap, but its use is, so far, limited to the fabrication of rather thick SOI films because of the absence of an etch stop. A special device has been developed to accurately control the thickness uniformity of the top silicon film [2.173]. It consists in a small-area, computer-controlled polishing tool with a continuous film thickness measurement feedback to the computer. Much more accurate are the techniques using, after initial grinding, a chemical etch-back procedure with etch stop(s). The etch stop is obtained by creating dopant concentration gradients at the surface (*i.e.* right next to the oxide layer used for bonding) of the top wafer. For instance, in the double etch-stop technique, a lightly doped wafer is used, and a P^{++} layer is created at its surface by ion implantation. Then, a lowly doped epitaxial layer is grown onto it. This epitaxial layer will be the SOI layer at the end of the process. After grinding, two chemical etch steps are used. The first one, an ethylenediamine-pyrocatechol-water etch [2.174], etches the substrate and stops on the P^{++} layer. Then, a 1:3:8 $HF:HNO_3:CH_3COOH$ etch is used to remove the P^{++} layer. The combined selectivity of the etches is better than 10,000:1 [1.170]. The final thickness uniformity of the SOI layer depends on the uniformity of the silicon thickness grown epitaxially, as well as on the uniformity of the P^{++} layer formation, but thickness standard deviations better than 12 nm can be obtained [1.170].

The crystal quality of the SOI material obtained by wafer bonding and etch-back is, in principle, as good as that of the starting silicon wafer. The dislocation density, however, is increased in the case where epi is used on a P^{++} buried layer as an etch stop. The dislocation density, however, is still below 10^3 cm^{-2}. The minority carrier lifetime is comparable to bulk silicon (≅20 μs). Circuits with complexity up to the 256k SRAM level have been demonstrated in SOI layers obtained by the bonding and etch-back technique (Si film thickness = 0.3 μm) [2.175].

2.10. What material for what application?

There are so many techniques for producing silicon-on-insulator material that one might have the impression that none of the proposed techniques can impose itself as a real winner. Of course, the description of the different SOI materials given in Chapter 1 is quite academic and it depicts, in a more or less historical perspective, most of the techniques developed to date.

Material	Defect density	Thin Si film thickness control	Minor. carrier lifetime	Channel mobility
SOS, as-grown	- -	0	- -	- -
SOS: DSPE, SPEAR	0	0	-	-
SOZ	- -	0	- -	- -
CaF2	- -	0	- -	- -
Laser	0	-	0	0
E-beam	0	-	0	0
ZMR	+	-	+	+
ELO	-	- -	-	-
LSPE	- -	+	- -	- -
FIPOS	+ +	+	+ +	+
Bur. Nitride	+	+	-	+
SIMOX	+	+	0/+	+
Bonding	+ +	- -	+ +	+
Bulk	+ +	N/A	+ +	+ +

Table 2.10.1: Comparison of some physical and electrical properties of the different SOI materials. Bulk silicon properties are indicated as a scale reference. ++= very good, +=good, 0=average, -=poor, --=very poor. Mobility figures do not include thin-film effects (enhanced mobility).

Some of the materials were mentioned regardless of their capability in meeting microelectronics quality specifications. The goal of the present Section is to sort out these techniques and suggest which material is best suited for a given application.

A comparison of some of the properties of the silicon films obtained by the different techniques is presented in Table 2.10.1. Some of these technique are still being developed at the stage of exploratory research (*e.g.*: epi on CaF_2, ELO, and LSPE), but are considered as possible contenders for the fabrication of 3D integrated circuits in the future. Some other techniques, on the other hand, have gained wide acceptance among the SOI community. These have been used to realize significant integrated circuits, some of which are produced commercially. Wafer bonding, SIMOX and ZMR materials are commercially available, and the manufacturers of these materials provide only wafers which strictly meet specifications concerning SOI film thickness, particulate density, defect density and contamination levels. Comparison of the different SOI materials in terms of maturity and commercial availability is given in Table 2.10.2.

Material	Country	Purpose	Commercially available?	Thin films ?	Maturity
SOS	US, J, Eu	Rad	Y (up to 5")	Y	+
SOZ	US, J, Eu		N	N	-
CaF2	J	3D	N	N	-
Laser	J, Eu	3D	N	Y	0
E-beam	US, J, Eu	3D	Y	Y	0
ZMR	US, Eu	Rad	Y (up to 6")	N	+
ELO	US, J, Eu	3D	N	N	-
LSPE	J	3D	N	Y	-
FIPOS	US, Eu	Rad	N	Y	+
Bur. Nitride	US, Eu	Rad	N	Y	-
SIMOX	US, J, Eu	Rad, ULSI	Y (up to 6")	Y	+
Bonding	US, J, Eu	Rad	Y (up to 6")	N	0

Table 2.10.2: Comparison of the different SOI materials. US=USA, J=Japan, Eu=Europe, Rad=rad-hard circuits, Y=yes, N=no, +=good, 0=average, -=poor.

One can distinguish three main application fields for SOI materials:

* *Rad-hard and VLSI CMOS applications.* In this class, SIMOX appears to be the dominant technology, although due to recent progresses, ZMR and wafer bonding techniques cannot be ruled out. Literature searches indicate that SIMOX is no longer only used as a replacement for SOS, but that it may become a major technology in the field of commercial CMOS application in a near future [2.176, 2.177].

* *"Smart power", power, high-voltage and bipolar applications.* Here, wafer bonding seems to be the best technique, owing to its low defect density, and the ability to produce thick silicon and buried oxide layers. ZMR is another possible contender.

* *Three-dimensional integration.* This field of application is still quite exploratory, but the laser and e-beam recrystallization techniques have proven to be most useful and sufficiently controlled for the fabrication of impressive 3D prototype circuits.

CHAPTER 3 - SOI Materials Characterization

Once Silicon-On-Insulator material has been produced, it is important to characterize it in terms of quality. The most critical parameters are the defect density, the thickness of both the top silicon layer and the buried insulator, the concentration of impurities, the carrier lifetime, and the quality of the silicon-insulator interfaces.

Parameter characterized	Characterization method	Destructive (D) or not (ND)	Physical / Electrical
Film thickness	Visible reflectance	ND	P
	Spectroscopic ellipsom.	ND	P
	IR reflectance	ND	P
	RBS	D	P
	Stylus profilemeter	D	P
	dVth/dVg2 technique	D	E
Si crystal quality	TEM	D	P
	RBS	D	P
	Chemical decoration	D	P
	UV reflectance	ND	P
Stress in Si layer	Raman spectroscopy	ND	P
Impurity content	SIMS	D	P
	Spark-source mass spectroscopy	D	P
Carrier lifetime	Surface photovoltage	ND	P
	Measurements on devices	D	E
Si-SiO2 interfaces	MOS capacitor	D	E

Table 3.1: Some SOI materials characterization techniques.

Some of the techniques used for characterization of SOI materials are very powerful, but destructive (*e.g.*: transmission electron microscopy - TEM). These are tools of choice for thorough examination of the materials, but can hardly be employed on a routine basis in a production environment. Some other techniques may be less sensitive, but are non destructive and provide information in a matter of seconds. These may be used to assess the quality of large amounts of SOI wafers.

Physical techniques, such as optical film thickness measurement techniques, can be employed on virgin SOI wafers, while some others need the fabrication of devices in the SOI material, and are, therefore, destructive by definition. We will now describe in more detail some of the characterization techniques used to assess the quality of SOI materials (Table 3.1).

3.1. Film thickness measurement

Accurate determination of the thickness of both the silicon overlayer and the buried insulator layer is essential for device processing as well as for the evaluation of parameters such as the threshold voltage. General-purpose film thickness measurement techniques can, of course, be utilized to evaluate SOI structures. These include the etching of steps in the material whose thickness is to be determined and the use of a stylus profilemeter to measure the height of the steps. Rutherford backscattering (RBS) can also be used to evaluate the thickness of a silicon film on SiO_2, but is not practical for the measurement of the buried oxide thickness, since oxygen atoms are lighter than silicon atoms. Cross-sectional transmission electron microscopy (XTEM) is probably the most powerful film thickness measurement technique, and atom lattice imaging can provide a built-in ruler for accurate distance measurements. All the above techniques are both destructive and time consuming. Routine inspection and thickness mapping of SOI wafers necessitate contactless (non-destructive) methods. Both spectroscopic reflectometry and spectroscopic ellipsometry satisfy this requirement. Electrical film thickness measurement can also be performed on manufactured fully-depleted devices. This measurement is based on the body effect (variation of front threshold voltage with back-gate bias) and can be used to verify that the thickness of the silicon film in which the devices are made is indeed what is expected from processing.

3.1.1. Spectroscopic reflectometry

The thickness of both the silicon overlayer and the buried oxide can be measured in a non-destructive way by measuring the reflectivity spectrum of an SOI wafer. Visible light (λ ranging from 400 to 800 nm) is usually utilized. The spectrum can then readily be compared with the theoretical reflectivity spectrum which can be calculated from the theory of thin-film optics. For SOI samples with well-defined Si-SiO_2 interfaces, a three-layer model can be used (silicon on oxide on silicon), while more complicated multilayer structures have to be employed if the interfaces are not well defined (in the case of as-implanted SIMOX material, for example [3.1]).

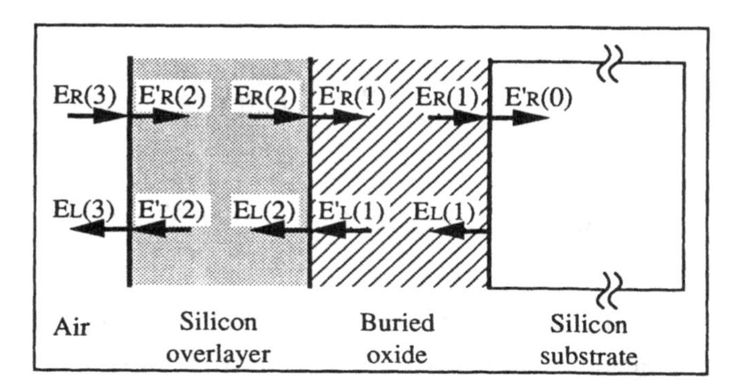

Figure 3.1.1: Multilayer structure and light-induced electric fields.

Let us consider the case of an SOI structure with sharp $Si\text{-}SiO_2$ interfaces (Figure 3.1.1) and use the following notation: the semi-infinite silicon substrate, the buried oxide, the silicon overlayer and the air ambient are numbered 0,1,2, and 3, respectively. The light-induced electric fields at the different interfaces between the n-th and the (n+1)-th layer are expressed by the following equation [3.2-3.4]:

$$\begin{pmatrix} E_R(n+1) \\ E_L(n+1) \end{pmatrix} = \mathbb{W}_{n+1,n} \begin{pmatrix} E'_R(n) \\ E'_L(n) \end{pmatrix} \quad (3.1.1)$$

where $E_R(n+1)$ and $E'_R(n)$ are the electric fields of the right-going wave at the left and right of the n-th interface, respectively, and $E_L(n+1)$ and $E'_L(n)$ are the fields of the left-going wave at the left and the right of the n-th interface, respectively. The interface matrix, $\mathbb{W}$, is given by:

$$\mathbb{W}_{(n+1,n)} = \frac{1}{2} \begin{pmatrix} 1 + \frac{y_n}{y_{n+1}} & 1 - \frac{y_n}{y_{n+1}} \\ 1 - \frac{y_n}{y_{n+1}} & 1 + \frac{y_n}{y_{n+1}} \end{pmatrix} \quad (3.1.2)$$

where y_n is the refractive index of the n-th layer. y_n is a complex number composed of a real (non-absorbing) and an imaginary (absorbing) part: $y_n=N_n+jK_n$. For the wavelengths under consideration, the refractive indices of dielectrics such as SiO_2 and Si_3N_4 can be considered as real and constant, while the refractive index of silicon is complex and wavelength dependent [3.5]. The phase change and the absorption occurring within the thickness of a layer (n-th layer) is expressed by the following relationship:

$$\begin{pmatrix} E'_R(n) \\ E'_L(n) \end{pmatrix} = \mathbb{\Phi}_n \begin{pmatrix} E_R(n) \\ E_L(n) \end{pmatrix} \qquad (3.1.3)$$

where the "phase matrix", $\mathbb{\Phi}$, is given by: $\mathbb{\Phi}_n = \begin{pmatrix} e^{j\Phi_n} & 0 \\ 0 & e^{-j\Phi_n} \end{pmatrix}$

where $\Phi_n = \frac{2\pi}{\lambda} y_n t_n$, t_n being the thickness of the n-th layer, and λ being the wavelength. Considering the full SOI structure, we finally obtain the following expression:

$$\begin{pmatrix} E_R(3) \\ E_L(3) \end{pmatrix} = \mathbb{W}_{3,2}\, \mathbb{\Phi}_2\, \mathbb{W}_{2,1}\, \mathbb{\Phi}_1\, \mathbb{W}_{1,0} \begin{pmatrix} E'_R(0) \\ E'_L(0) \end{pmatrix} \qquad (3.1.4)$$

We will note that the silicon substrate is considered as semi-infinite. Therefore, there is no returning wave from the substrate in the above equation ($E'_L(0) = 0$). The reflectivity of the structure is given by the ratio of reflected power to incident power [3.3]:

$$R(\lambda) \equiv \frac{[E_L(3)]^2}{[E_R(3)]^2} \qquad (3.1.5)$$

The reflectivity is expressed in % of incident power, and $E'_R(0)$ is easily eliminated from equation (3.1.5). Indeed, $E'_R(0)$ can be viewed as a reference value and can be considered to be equal to unity ($E'_R(0)=1$). Equation (3.1.4) then yields the values for $E_R(3)$ and $E_L(3)$ from which the ratio $R(\lambda)$ can be computed for each wavelength. The above relationships were established for normal incidence of the light on the sample. Angles of incidence up to ...15 degrees... can, however, be used without introducing any significant error in the measurement. An example of reflectivity spectrum of a SIMOX wafer annealed at high temperature (*i.e.*: with sharp Si-SiO_2 interfaces) is given in Figure 3.1.2. The determination of the thickness of both the silicon film and the buried layer is based on the comparison between the measured spectrum and the calculated one. The film thickness values can be obtained iteratively through minimizing the difference between

$R_{measured}$ and $R_{calculated}$ at selected wavelengths. This iterative technique is unfortunately rather time consuming.

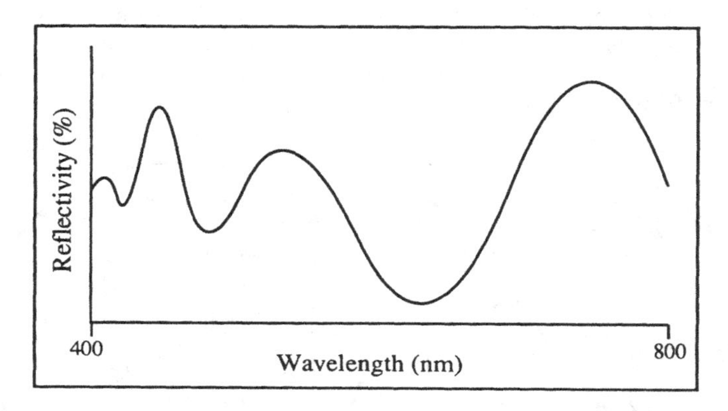

Figure 3.1.2: Example of reflectivity spectrum of a SIMOX sample.

There exists, however, another comparison technique which is significantly faster. It is based on the following observation: the reflectivity spectra present a succession of extrema (minima and maxima), and that the position of these extrema is a function of the thickness of both the silicon overlayer and the buried oxide.

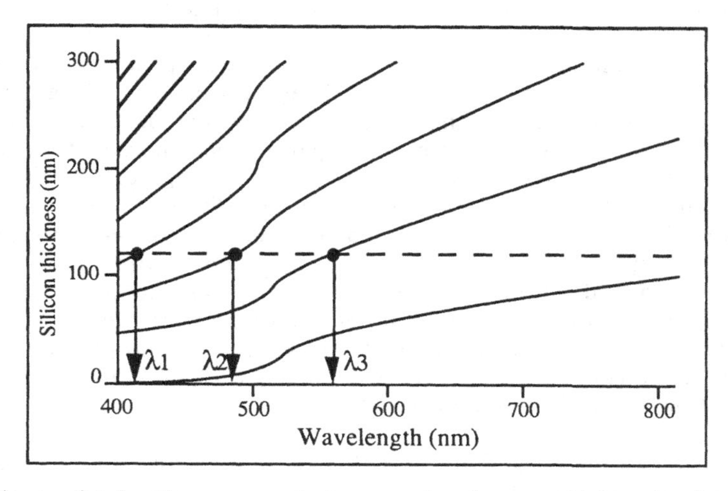

Figure 3.1.3: Nomogram of the wavelengths at which reflectivity minima occur (x-axis) as a function of silicon overlayer thickness (y-axis). The buried oxide thickness is 352 nm. For a silicon thickness of 120 nm, for instance, three reflectivity minima are found (at $\lambda = \lambda_1, \lambda_2$, and λ_3).

Figure 3.1.3 presents a nomogram of the position of the reflectivity minima (or absorbance maxima in ref. [3.6]) of a SIMOX structure having a 352 nm-thick buried oxide, as a function of the silicon overlayer thickness. It can be observed that both the number of minima and the position (the wavelengths at which they occur) is a function of the silicon film thickness).

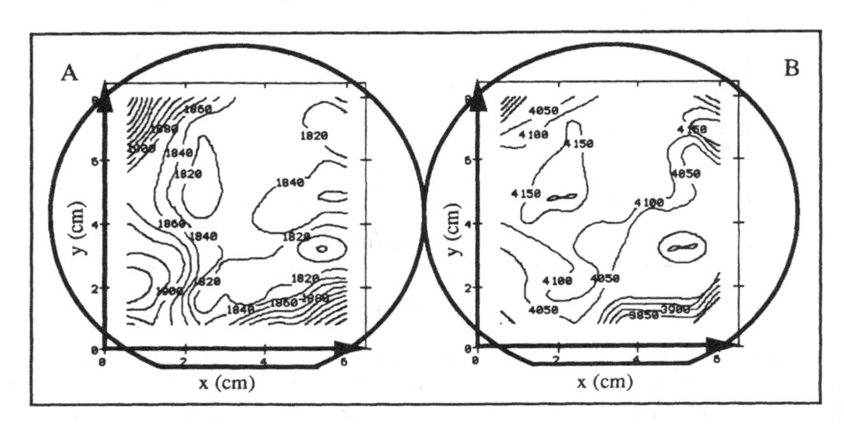

Figure 3.1.4: Mapping of the thickness of the silicon overlayer (A) and the buried oxide (B) of a 125 mm-diameter SIMOX wafer. Thickness is indicated in angströms (Å). (1Å = 0.1 nm).

A database containing the position of the reflectivity minima for a wide range of silicon and buried oxide thicknesses can readily be computed and stored in a computer memory. Accurate estimation of film thickness can then be obtained by merely comparing the measured reflectivity minima with the values stored in the database. Figure 3.1.4 shows an example of thickness mapping made on a SIMOX wafer using this technique [3.7]. Measurements based on reflectivity measurements in the visible spectrum are efficient for silicon films thinner than ...500nm... Indeed, thicker SOI layers are no longer sufficiently transparent to visible light to allow for optical interferences to produce clear, well-defined reflectivity extrema and, in addition, thick films produce a large number of extrema, which slows down the interpretation of the measured spectra.

In the case of thick SOI films, longer wavelengths (infrared, $1\ \mu m < \lambda < 5\ \mu m$) can be used. These can penetrate through the silicon film and produce interference patterns which can be used to determine the film thickness of the SOI structure. Silicon-on insulator can be modelled as a double layer structure on a substrate. Under normal incidence conditions, the reflectance can be expressed by [3.8, 3.9]:

$$R(\lambda) = \frac{A + f(\theta_1,\theta_2)}{F + f(\theta_1,\theta_2)}$$

where

$$f(\theta_1,\theta_2) = a_1\cos 2\theta_1 + a_2\cos 2\theta_2 + a_3\cos 2(\theta_1+\theta_2) + a_4\cos 2(\theta_1-\theta_2)$$

$$A=r_1^2+r_2^2+r_3^2+r_1^2r_2^2r_3^2$$

$$a_1=2r_1r_2(1+r_3^2)$$

$$a_2=2r_2r_3(1+r_1^2)$$

$$a_3=2r_1r_3$$

$$a_4=2r_1r_2^2r_3$$

$$F=1+r_1^2r_2^2+r_1^2r_3^2+r_2^2r_3^2$$

The *r's* are the individual Fresnel reflectances of the three interfaces (air-silicon overlayer, silicon overlayer-buried oxide, and buried oxide-silicon substrate) and are given by: $r_j = \frac{n_{j-1}-n_j}{n_{j-1}+n_j}$, j=1,2,3. θ_1 and θ_2 are the phase thicknesses of the silicon and the buried oxide layers, respectively, and are given by: $\theta_j = \frac{2\pi}{\lambda} n_j t_j$, (j=1,2). The n_j are the real refractive indices of the different layers (being transparent at the wavelengths under consideration, both silicon and silicon dioxide have real refractive indices). The t_j are the layer thicknesses. By expanding the denominator of (3.1.6) in a power series, one finds:

$$R(\lambda) = \frac{A + f(\theta_1,\theta_2)}{F + f(\theta_1,\theta_2)} = \frac{A}{F}\,\frac{1+f(\theta_1,\theta_2)/A}{1+f(\theta_1,\theta_2)/F}$$

Using $\frac{1}{1+f(\theta_1,\theta_2)/F} \cong \left(1-(f(\theta_1,\theta_2)/F)+(f(\theta_1,\theta_2)/F)^2 +...\right)$ for $(f(\theta_1,\theta_2)/F)^2 < 1$, one obtains:

$$R(\lambda) \cong \frac{A}{F}\,(1+f(\theta_1,\theta_2)/A)\left(1-f(\theta_1,\theta_2)/F+(f(\theta_1,\theta_2)/F)^2 - ...\right)$$

or

$$R(\lambda) \cong \frac{A}{F}\left(1 + (1/A-1/F)\, f(\theta_1,\theta_2) + (1/F^2-1/AF)\, f^2(\theta_1,\theta_2) + ...\right)$$

or, more generally,

$$R(\lambda) = R_o \sum_{n=0}^{\infty} k_n\, f^n(\theta_1,\theta_2)$$

where the k_n are functions of the individual Fresnel reflectances of the three interfaces. Let us consider, for instance, the first term in the expression for $f(\theta_1,\theta_2)$: $a_1 \cos 2\theta_1 = a_1 \cos\left(2\pi \cdot 2n_1t_1 \cdot \frac{1}{\lambda}\right)$ where $2n_1t_1$ is the optical thickness for a wave traversing twice the layer of refractive index n_1. As expressed above, the optical thickness is analogous to frequency, and the inverse wavelength is analogous to time in the parlance of conventional Fourier analysis. The reflectance spectrum contains several "frequency" components corresponding to $2n_1t_1$, $2n_2t_2$, $2(n_1t_1+n_2t_2)$, $2(n_1t_1-n_2t_2)$, and multiples thereof. By applying the so-called "bilinear transform" $B\{R(\lambda)\} = \frac{1+R(\lambda)}{1-R(\lambda)}$, the number of frequency components can be reduced dramatically [3.9]. The effects of the additional frequency modulation as a result of the wavelength dependence of the refractive indices are minimized by analyzing only data in the region between approximately 1 μm and 5 μm where the refractive indexes of both Si and SiO_2 are constant. Under these conditions, only three frequency components are dominant, and the Fourier analysis of $B\{R(\lambda)\}$ yields peaks at the following wave numbers: $\nu_{Si}=2n_1t_1$, $\nu_{SiO2}=2n_2t_2$, and $\nu_{total(Si+SiO2)}=2(n_1t_1+n_2t_2)$, which are the optical thicknesses of the silicon layer, the buried oxide layer, and the Si/SiO_2 structure, respectively. Conversion from wave numbers at which peaks occur (ν_{Si} and ν_{SiO2}) into layer thicknesses can be done using the following expressions: $t_{Si}=\frac{10^7\nu_{Si}}{2n_{Si}}$ and $t_{SiO2}=\frac{10^7\nu_{SiO2}}{2n_{SiO2}}$, where ν is measured in cm and t in nm. This technique is complementary to the visible reflectance technique since it can be used to measure the thickness of relatively thick SOI layers, owing to the longer wavelengths used here (infrared), but it can also be used for thinner films (down to 100 nm) [3.9].

3.1.2. Spectroscopic ellipsometry

Ellipsometry is based on the measurement of the change of polarization of a light beam reflected by a sample. In order to maximize the sensitivity of the measurement, ellipsometry is usually carried out at large incidence angles (75°, close to the Brewster angle). The change of polarization can be derived from equations (3.1.1)-(3.1.4) where corrections for non-normal incidence have to be introduced. The complete theory of ellipsometry is quite complicated and is outside the scope of this book. The interested reader can, however, refer to [3.10] for more information. Classical ellipsometry is performed at a single wavelength (usually emitted by an He-Ne laser). After passing through a polarizing filter, the beam is reflected on the sample and is directed into an analyzer, composed of a rotating polarization filter and a photodiode. The polarization direction of the reflected beam is given by the rotation angle of the latter filter when the amplitude of the light collected by the photodiode is maximum. In spectroscopic ellipsometry, the same analysis is repeated for a large number of wavelengths within the visible spectrum and the near ultraviolet. Typically, about a hundred equidistant measurements are performed at wavelengths ranging between 300 and 850 nm.

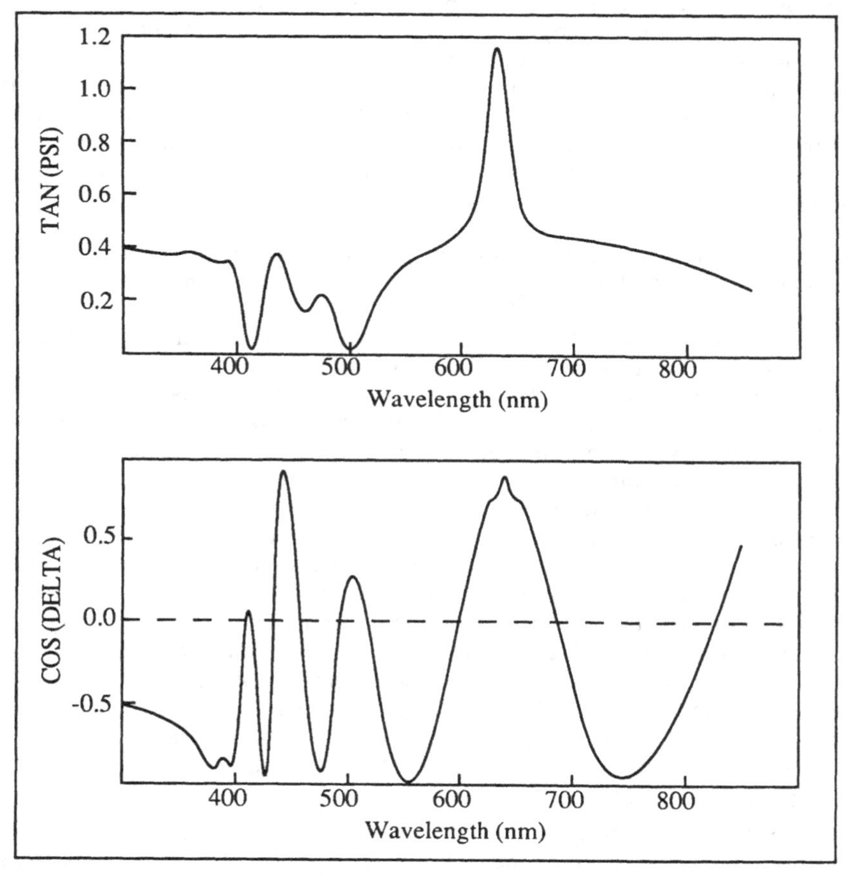

Figure 3.1.5: Example of the wavelength dependence of the ellipsometry parameters $\tan\Psi$ and $\cos\delta$ in a SIMOX structure [3.7].

The output of the measurement consists into the spectra of $\tan\Psi$ and $\cos\delta$, which angles are defined, at each wavelength, by the relationship:

$$\tan\Psi \; e^{j\delta} = \frac{r_p}{r_s}$$

where r_p and r_s are the complex reflection coefficients *r* for light polarized parallel (p), and perpendicular (s - from *senkrecht*, in German) to the plane of incidence. An example of $\tan\Psi$ and $\cos\delta$ spectra is given in Figure 3.1.5. The interpretation of the measured spectra is based on a simulation and regression program which minimizes the difference between the measured data and spectra calculated using thin-film optics theory. Spectroscopic ellipsometry spectra contain much more information than spectroscopic reflectometry spectra. Its sensitivity is such that it can be used to measure complicated multilayer structures such as imperfect SIMOX structures which contain silicon inclusions within the buried oxide and oxide precipitates within the silicon overlayer. It can even measure the thickness of a native oxide layer on top of an SOI wafer [3.7].

Spectroscopic ellipsometry is not an analytical technique which stands of its own. Indeed, parameters such as the number of layers to be taken into account as well as their composition have to be fed into the simulator which will endeavor to tune each parameter in order to reproduce the measured data. For example, a SIMOX structure with oxide precipitates at the bottom of the silicon overlayer and silicon precipitates at the bottom of the buried oxide will need the following input parameters: estimation of the thickness of the following layers: the native oxide, the silicon overlayer without precipitates, the silicon overlayer with precipitates, the buried oxide above the silicon precipitates, the buried oxide with silicon precipitates, and the oxide below the precipitates. An estimation of the composition of the mixed layers (*e.g.*: silicon precipitates in oxide) must also be given. The refractive indices of mixed layers are calculated using the Bruggeman approximation (or effective medium approximation) [3.11]. This approximation assume an homogeneous isotropic mixture of spheres of both components (*e.g.*: Si and SiO_2) with varying radius in order to obtain complete filling of the layer. In this approximation, the mixed layer is isotropic, and its dielectric constant, ε, is given by:
$\varepsilon = 0.25\left(E + \sqrt{E^2 + 8\varepsilon_1\varepsilon_2}\right)$ with $E = (3c-1)(\varepsilon_2-\varepsilon_1) + \varepsilon_1$, where c is the fraction of component 2 within the mixed layer, and ε_1 and ε_2 are the dielectric constants of component 1 and component 2, respectively. ε, ε_1, ε_2 and are complex numbers from which the expression of the refractive index $n=\sqrt{\varepsilon}$ can be extracted. The regression analysis during which measured and calculated spectra are compared, endeavors to minimize

the error function $G = \sum_{i=1}^{n} [(D_i^c - D_i^m)^2 + (W_i^c - W_i^m)^2]$ where n is the number of different wavelengths at which the measurement is performed. The superscripts *m* and *c* stand for "measured" and "calculated", respectively. The parameters W and D are given by the following expressions: $W_i = \frac{\tan^2\Psi_i - 1}{\tan^2\Psi_i + 1}$ and $D_i = \frac{2\cos\delta_i \tan\Psi_i}{\tan^2\Psi_i + 1}$ or $D_i = \cos\delta_i$ if $\tan\Psi_i$ is very small [3.7].

Spectroscopic ellipsometry is a very sensitive measurement technique. When the layers and interfaces of the SOI structure are well defined, a three-layer model can be used (native oxide/silicon/buried insulator), and convergence of the thickness-finding algorithm can be obtained in a matter of minutes. If the composition of the interfaces is not well defined (*e.g.*: as-implanted or low-temperature annealed SIMOX), such that mixed layers have to be simulated, many parameters have to be optimized simultaneously, and convergence can take up to several hours.

3.1.3. Electrical thickness measurement

The silicon film thickness, t_{si}, is an important parameter in thin-film, fully-depleted SOI MOSFETs. Indeed, t_{si} influences all the electrical parameters of thin-film devices (threshold voltage, drain saturation voltage, subthreshold slope, ...). Therefore, it is of interest to measure the thickness of the silicon film after device processing for debugging purposes and to check whether threshold voltage non-uniformities, for example, are due to film thickness variations or not, and whether the final targeted silicon thickness has been reached after device processing.

The dependence of the front threshold voltage, V_{th1}, of a fully-depleted n-channel SOI MOSFET on back-gate voltage, V_{G2}, is given by equation (5.3.13) of Section 5.3 (body effect):

$$\frac{dV_{th1}}{dV_{G2}} = -\frac{C_{si}\,C_{ox2}}{C_{ox1}\,(C_{si} + C_{ox2})} = \frac{-\varepsilon_{si}\,C_{ox2}}{C_{ox1}\,(t_{si}\,C_{ox2} + \varepsilon_{si})}$$

from which one can easily derive the following relationship [3.12]:

$$t_{si} = -\frac{1}{C_{ox2}}\left(\frac{\varepsilon_{si}\,C_{ox2}}{C_{ox1}} - \varepsilon_{si}\right)\left(\frac{dV_{th1}}{dV_{G2}}\right)^{-1}$$

where C_{ox1}, C_{ox2}, and ε_{si} are the gate oxide capacitance, the buried oxide capacitance, and the silicon permittivity, respectively. It is important to note that the above expression is independent of the dopant concentration in the silicon film, as long as the device is fully depleted. This measurement technique assumes that C_{ox1} and C_{ox2} (or t_{ox1} and t_{ox2}) are known. The value of t_{ox1} can be obtained from an independent measurement made on bulk silicon wafers undergoing the same gate oxidation gate process as the SOI wafers, and the buried oxide thickness (which is not affected by device processing) can be measured by spectroscopic reflectometry mapping techniques before processing. It is also worth noting that the dV_{th1}/dV_{g2} measurement must be carried out for back-gate voltages for which the back-gate interface is depleted(*i.e.* neither inverted nor accumulated) (refer to Figure 5.3.4).

3.2. Crystal quality

Although all SOI material producing techniques aim at the fabrication of a perfect single-crystal silicon layer, they never completely achieve their goal, and crystal imperfections are found within the silicon. This Section describes some of the various techniques used to assess the quality of the silicon layers (crystal orientation, crystallinity, and crystal defects).

3.2.1. Crystal orientation

All SOI techniques are designed to produce silicon films with (100) normal orientation. This orientation is automatically obtained when the silicon film is produced by separation of a superficial silicon layer from a (100) silicon substrate by the formation of a buried insulator (SIMOX, SIMNI, FIPOS or wafer bonding) or when the silicon film is epitaxially grown from a single-crystal substrate having normal lattice parameters equal or close to those of (100) silicon (SOS, ELO, LSPE,...). In those cases where the silicon film is recrystallized from the melt over an amorphous insulator (laser and e-beam recrystallization, ZMR), the control of the crystal orientation is more difficult, and substantial deviation from the (100) orientation can be observed. Furthermore, even if the normal orientation is (100), in-plane misorientation can occur. The different single-crystals are then connected by subgrain boundaries.

The crystal orientation can be determined by classical techniques such as X-ray diffraction and electron diffraction in a TEM. More frequently, the electron channeling pattern (ECP) technique (or: pseudo-Kikuchi technique) is employed since it can be carried out in an SEM without any special preparation of the sample. In this technique, a stationary, defocussed electron beam is incident onto the sample. The reflection of the electrons depends on the local crystal orientation relative to the incident beam (*i.e.*: it depends on the degree of channeling of the incident electrons in the silicon). The reflected electrons form a pattern which is indicative of the normal crystal orientation (a cross-shaped pattern is obtained for (100) silicon). The direction of the arms of the cross pattern can be used to determine the in-plane orientation, and a distortion of the pattern indicates a spatial rotation of the crystal axes [3.13].

One of the most popular methods used to assess the crystal orientation of SOI films is the etch-pit grid technique [3.14]. This technique has been widely used to optimize the ZMR process. In this technique, a layer of oxide is grown or deposited on the silicon layer. Using a mask step and HF etch, circular holes are opened in the oxide layers. The holes have a 2-3 μm diameter and are repeated across the entire sample in a grid array configuration, with a pitch of 20 μm, typically. The resist is then stripped, and the holes in the oxide are used as a mask for silicon etching. The silicon is etched in a KOH solution (a mixture of 250g KOH, 800 ml deionized water and 200 ml isopropyl alcohol was used in ref. [3.14]). The KOH solution etches silicon much more rapidly in the <100> direction than in the <111> direction. As a result, a pattern having the shape of a section of a pyramid is etched in the silicon. This pattern has a square shape if the normal orientation is (100), and the sides of the pits are <111> planes. The pit diagonals indicate [100] directions (Figure 3.2.1).

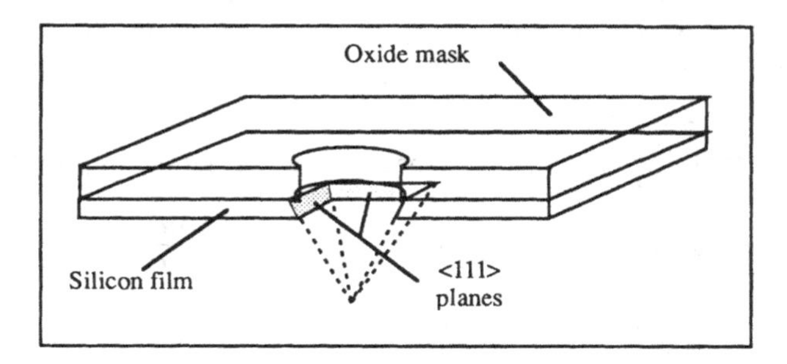

Figure 3.2.1: The etch-pit grid technique.

If the film orientation is not (100), distorted polyhedral patterns are produced by the intersection of the silicon film plane with the <111> silicon planes octahedron. The change of shape between two adjacent patterns indicates the presence of grain boundaries or the rotation of the crystalline axes. A misalignment of the diagonal directions between two patterns without modification of shape is indicative of the presence of subgrain boundaries.

3.2.2. Degree of crystallinity

In those cases where the silicon overlayer is damaged during the SOI formation process (SIMOX, SIMNI), and when epitaxial growth of silicon is performed, it is sometimes interesting to check the single-crystallinity of the formed layer. Two major techniques may help us to achieve that goal: Rutherford backscattering (RBS) and UV reflectance.

RBS is a destructive technique based on the impingement of light ions (usually He^+) with mass M_1 on a sample consisting of atoms with mass M_2. The ions are accelerated to an energy E_0 (*e.g.*: 2 MeV) before reaching the target. These ions loose energy through nuclear and electronic interactions with the target atoms. Most of them will come to rest in the target, but a small fraction of these light ions will be backscattered over an angle Θ with an energy $E_1 = K E_0$, with K being the kinetic factor [3.15]:

$$K = \frac{\sqrt{M_2^2 - M_1^2 \sin\Theta} + M_1 \cos\Theta}{M_1 + M_2}$$

The detection of the energy of the recoiling atoms can thus be used to determine the mass M_2 of the target atoms. The probability for an elastic collision to occur and to result in a scattering event at a certain angle Θ is expressed by the differential scattering cross-section:

$\frac{d\sigma}{d\Omega} \propto \left(\frac{Z_1 Z_2 q^2}{4 E_0}\right)^2 \frac{1}{\sin^4\Theta}$, with Ω being the detector angle. The average scattering cross-section is then given by: $\sigma = \frac{1}{\Omega}\int_{\Omega}\frac{d\sigma}{d\Omega} d\Omega$. The stopping cross-section $\varepsilon = \frac{1}{N}\frac{dE}{dx}(E)$ finally accounts for the energy loss of the particle penetrating the target due to electronic collisions or to small-angle collisions with nuclei; x is the depth below the target surface, and N is the volumic density of the target atoms. The detected signal is processed by a multichannel analyzer The output of a measurement session is a series of counts, called the backscattering yield, in every channel. To interpret the measured data, one has to convert the channel numbers into an energy scale and, therefore, to determine the energy interval ε corresponding to a channel. Actually, the useful information is the depth scale, and E must be correlated with a slab i of thickness τ_i at a depth x_i. It can be shown [3.15] that the energy difference between a particle backscattered at the surface ($E=KE_0$) and another one backscattered at a depth x and emerging from the target ($E=E_x$) is given by:

$$\Delta E = KE_0 - E_x = \left(\frac{K}{\cos\Theta_1} \left.\frac{dE}{dx}\right|_{in} + \frac{K}{\cos\Theta_2} \left.\frac{dE}{dx}\right|_{out} \right) x = [\varepsilon]\, N\, x$$

where Θ_1 and Θ_2 are the angles (with respect to normal) of the track of the particle before and after scattering in the target. The latter relationship assumes that dE/dx is constant along each path taken by the particle. This assumption yields a linear relationship between the energy difference, ΔE, and the depth at which scattering occurs. $[\varepsilon]$ is the stopping cross-section factor. It is worthwhile noting that the analysis of RBS spectra can give information on the composition of compound materials (such as SiO_2 or, more generally, Si_xO_y).

An RBS spectrum measurement can be performed in two different ways: a crystal direction can be parallel to the incident ion beam, or it can be randomly oriented. In the former case, the ions can penetrate deeper in the crystal by channeling through the lattice, and an "aligned spectrum" is obtained. If the sample is amorphous or randomly oriented, no channeling can occur, and a "non-aligned" spectrum is obtained. Aligned spectra have lower backscattering yield because the ions penetrate deeper in the sample and have a lower probability to escape out of it after a collision. Similarly, the presence of crystalline imperfections (point defects, impurities,...) increases the backscattering yield of a crystalline, aligned target. The minimum backscattering yield, χ_{min}, is, therefore, a measure of the lattice disorder. The lower χ_{min}, the better the crystallinity. Single-crystal (100) bulk silicon has a value of χ_{min} of 3-4%.

Figure 3.2.2 shows typical RBS spectra obtained from a SIMOX sample. The most useful information come from the layers nearest to the surface, where the energy of the backscattered ions is highest. Part (a) of Figure 3.2.2 corresponds to the silicon overlayer. The non-aligned spectrum shows a high yield and provides information on the thickness of the silicon layer. The aligned spectrum gives information on its crystal quality, through χ_{min}. Part (b) corresponds to the buried oxide. Both aligned and non-aligned spectra contain information on the thickness of the layer. The non-aligned spectrum provides information on the composition of the layer (the ratio of oxygen to silicon atoms). The yield of the non-aligned spectrum is lower for SiO_2 than for silicon because SiO_2 is lighter than Si. The aligned spectrum has a higher yield for SiO_2 than for Si because SiO_2 is amorphous, and no channeling can take place in the buried layer. Part (c) corresponds to the silicon substrate, and the peak which can be observed in part (d) is caused by influence of the buried oxide on the ions on their way back to the surface after a collision in the silicon substrate.

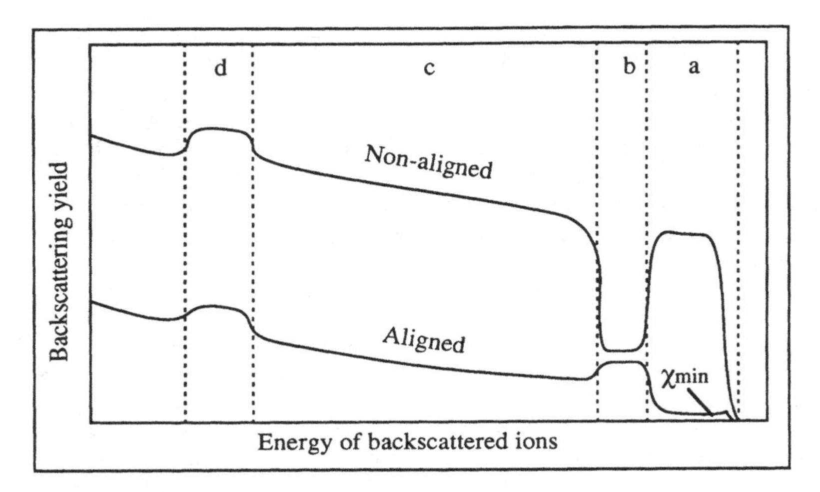

Figure 3.2.2: Aligned and non-aligned RBS spectra of a SIMOX wafer. From right to left: signal from the surface and the silicon overlayer (a), the buried oxide (b), the substrate (c), and influence of the buried oxide on the ions backscattered from the substrate (d).

UV reflectance is another technique which can be used to assess the crystallinity of SOI samples. Contrarily to RBS, it is non-destructive. It has extensively been used to characterize SOS wafers, and both the microtwin density in the film and the fabrication yield of SOS circuits can be correlated to UV reflectance parameters [3.16]. The quality of the silicon overlayer of SIMOX wafers can also be assessed using the UV reflectance technique [3.17]. In the case of SOS, the methods employs a measurement of the reflectance at λ=280 nm and, if necessary, a second measurement at λ=400 nm for reference [3.16]. There are two prominent maxima in the UV reflectance spectrum of single-crystal silicon, at λ=280 nm and λ=365 nm. They are caused by the optical interband transitions at the X point and along the Γ-L axis of the Brillouin zone, respectively. At short wavelengths, in particular at 280 nm, the reflectance is largely determined by the high value of the absorption coefficient $K > 10^6$ cm^{-1} corresponding to a penetration depth of less than 10 nm. Imperfect crystallinity in the near-surface region cause a broadening of the reflectance peak and a reduction of its maximum value [3.17]. In the case of SIMOX wafers, more wavelengths have to be taken into consideration in order to obtain useful information about the quality of the silicon overlayer. UV measurement of SIMOX wafers has been show to provide information on three morphological features of the material. Firstly, the overall reflectivity reduction (compared to a bulk silicon reference sample) is related to the presence of contamination in the film. This contamination can be due to the presence of carbon or SiO_x. Secondly, Rayleigh scattering caused by surface roughness shows a decrease of the reflectivity as a function of $B\lambda^{-4}$ for the shortest wavelengths (200 nm $< \lambda <$ 250 nm), where B is a constant depending on the rms roughness of the surface. Thirdly, some amorphization of the silicon overlayer reduces the intensity of the reflectance peaks at 280 and 367 nm. The sharpness and intensity of these maxima gives a measure of the degree of crystallinity within the specimen. Hence, semi-quantitative information about contamination, surface roughness and crystallinity can be obtained from UV reflectance measurements [3.17].

3.2.3. Crystal defects

Transmission electron microscopy (TEM) is one most powerful technique for the analysis of crystal defects. It is nevertheless limited by the size of the samples which can be analyzed. In cross-section TEM (XTEM), the dimensions of a sample which can be observed at once are approximately limited to a width of 20 μm and a depth of 0.7 μm. This means that the maximum observable area is on the order of 10^{-7} cm^{-2} and, consequently, that the minimum measurable defect density is approximately 10^7 defects/cm^2. Plane-view TEM allows one to observe larger sample areas. Areas of the size of the sample holder (7 mm^2 grid) can indeed be analyzed. In practice, it becomes difficult to observe dislocations with a magnification lower than 10,000 and it is more realistic to consider that an observation session yields 10 micrographs, each with a 10.000X magnification. In that case, the observed area is equal to 10^{-5} cm^2, and the minimum observable defect density is 10^5 defects/cm^{-2}. TEM observation often necessitates a lengthy and delicate sample preparation. Defect decoration techniques, combined with optical microscope observation are, therefore, preferred to TEM if the nature of the sample and the defects allow it. The main defects found in SOI materials are listed in Table 3.2.1. (Micro)twins and stacking faults are dominant in heteroepitaxial materials, while (sub)grain boundaries are found in silicon films recrystallized from the melt. Dislocations are the main crystal defect in SIMOX, FIPOS and material produced by wafer bonding.

Material	Type of defect	Concentration
SOS	Microwins, stacking faults	H
SOZ	Microwins, stacking faults	H
CaF2	Microwins, stacking faults	H
Laser	Grain boundaries, stacking faults	H, but localized
E-beam	Grain boundaries, stacking faults	H
ZMR	Subgrain boundaries, dislocations	M
ELO	Stacking faults, dislocations	H
LSPE	Dislocations, stacking faults	H
FIPOS	Dislocations	L
SIMOX	Dislocations	L
Bur. Nitride	Dislocations	M
Bonding	Dislocations	L

Table 3.2.1: Main types of crystal defects present in the different SOI materials. H="high", M="medium", L="low".

The most common etch mixtures used for SOI defect decoration are listed in Table 3.2.2. Except for the last one (electrochemical etch), all are based on the mixture of HF with an oxidizing agent (CrO_3, $K_2Cr_2O_7$ or HNO_3). Defect decoration stems on the preferential etch (higher etch rate) of the defects with respect to silicon.

Etch name	Composition	Ref.
Dash	HF:HNO3:CH3COOH 1:3:10	3.18
Schimmel	HF:1M CRO3 2:1	3,19
Secco	HF:0.15M K2Cr2O7 2:1	3. 20
Stirl	HF:5M CrO3 1:1	3.21
Wright	60ml HF:30ml HNO3:30ml 5M CrO3: 2 grams Cu(NO3)2:60ml CH3COOH:60ml H2O	3.22
Electrochemical	5% wt HF	3.23

Table 3.2.2: Decoration etchants used to reveal defects in SOI films.

Decoration is most effective for high-disorder defects such as grain and subgrain boundaries. The etch rate of silicon is approximately 1 μm/min for most mixtures (Dash, Secco, Sirtl, Wright) - (the Schimmel etch rate is substantially lower). The decoration of dislocations in thin-film SOI material is almost impossible using classical etch mixtures, since all the silicon is removed before efficient decoration of the defects is achieved. Therefore, a new decoration technique, based on the electrochemical etching of silicon in diluted (5%) HF has been developed to reveal crystal defects in thin SOI films without etching off the silicon overlayer itself [3.23]. This technique necessitates the use of n-type ($N_d \cong 10^{15}$ cm^{-3}) doped silicon overlayers, and an ohmic contact must be provided to both the front side of the sample (*i.e.* to the SOI layer) and to the back of it. Electrochemical etching is performed for 10-30 minutes in 5 wt.% HF using a three-electrode configuration with the silicon controlled at +3 volts *vs.* a Cu/CuF_2 reference electrode. This decoration technique does nor etch defect-free silicon. Defects such as dislocations, metal contamination-related defects, and oxidation-induced stacking faults (OSF) produce pits in the silicon film during this electrochemical etch procedure, and optical microscopy is used to observe and count the pits after decoration [3.23].

The incidence of a crystal defect on the electrical properties of a device depends on the nature and the geometry of the defect. We will now briefly describe the influence of some major crystal defects on the electrical properties of MOS devices made in SOI films.

3.2.3.1. (Sub)grain boundaries

A grain boundary (GB) is found when two grains with different crystal orientations meet. GBs are typical of laser and e-beam recrystallized SOI materials. The presence of a grain boundary in the channel region of a device gives rise to different effects, depending on the location of the boundary. If a GB runs from source to drain (the GB is parallel to the current direction), it acts as an enhanced-diffusion path for source and drain dopant impurities during S&D reoxidation. As a consequence, source-to-drain leakage is observed in transistors with relatively short channel lengths (Figure 3.2.3.A) [3.24].

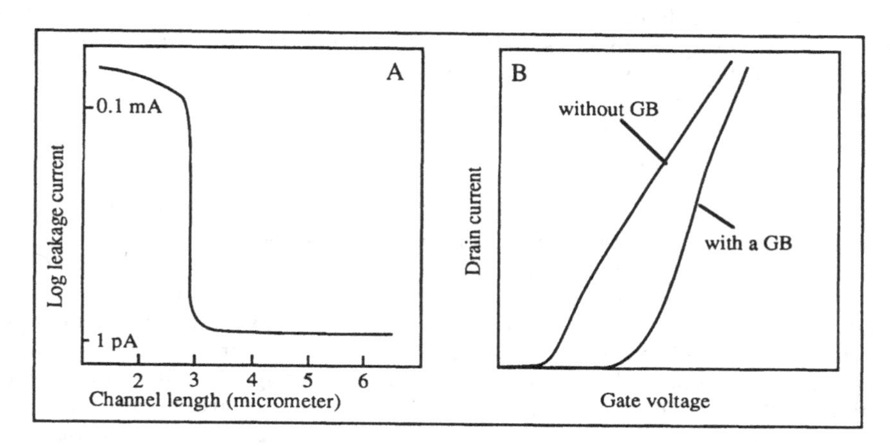

Figure 3.2.3: Influence of a grain boundary located in the channel area of SOI MOSFETs. The GB is either parallel (A) or perpendicular (B) to the current flow direction.

If the grain boundary is perpendicular to the current flow in the channel, the high density of interface traps in the grain boundary (N_{it}≅several 10^{12} cm^{-2}) retards the appearance of an inversion layer as gate voltage is increased. As a result, a dramatic increase of threshold voltage is observed (Figure 3.2.3.B). A significant reduction of the mobility in the inversion layer is also observed [3.25, 3.26].

Subgrain boundaries are found in ZMR material and are caused by the presence of two adjacent crystals with same normal crystal orientation, but with slightly different in-plane crystalline directions. A subgrain boundary (SGB), which can be viewed as a mere dislocation network, has much less electrical activity than a grain boundary. SGBs do not appear to either enhance dopant diffusion or to significantly degrade threshold voltage or mobility characteristics.

3.2.3.2. Dislocations

Dislocations are the main defect found in many SOI materials. In the case of SIMOX, the dislocations are threading dislocations running vertically from the Si/buried oxide interface up to the surface of the silicon overlayer. The presence of such dislocations may pose yield and reliability hazard problems. Indeed, metallic impurities readily diffuse to dislocations upon annealing, and dislocations decorated with heavy metal impurities can cause weak points in gate oxides, so that low breakdown voltage is observed. A 1987 study shows, nevertheless, that the integrity of gate oxides grown on SIMOX is comparable to that of oxides grown on bulk silicon [3.27]. In addition, SIMOX technology has since then brought about steady improvement of both the dislocation density and the heavy metal contamination level.

3.3. Silicon film contamination

Several types of contaminants can be introduced in the silicon overlayer during the SOI fabrication process. In the case of SIMOX, for instance, heavy metals and carbon can be sputtered from implanter parts and implanted into the wafer. Most forms of contamination can be analyzed by SIMS (secondary ion mass spectroscopy), but this technique is destructive and necessitates a rather heavy investment. Heavy metal contamination monitoring can be carried out using the SPV technique, which will be described in the next Section. Carbon and oxygen contamination are also important in SOI layers, and can contribute to stress and defect generation as well as to a decrease of the silicon overlayer resistivity.

3.3.1. Carbon contamination

Carbon contamination of SIMOX material can occur during the oxygen implantation step, most likely by interaction between the oxygen beam and either graphite implanter parts or the residual hydrocarbons present in the vacuum of the accelerator column. Carbon is known to form nucleation sites for oxygen precipitation and dislocations, such that a high ($>10^{18}$ cm^{-3}) carbon concentration can contribute to the generation of large amounts of dislocations in the silicon overlayer upon annealing of the SIMOX structure. Strong correlation has been established between the carbon concentration (measured by SIMS) and the dislocation density in the silicon overlayer of SIMOX wafers (Figure 3.3.1) [3.28].

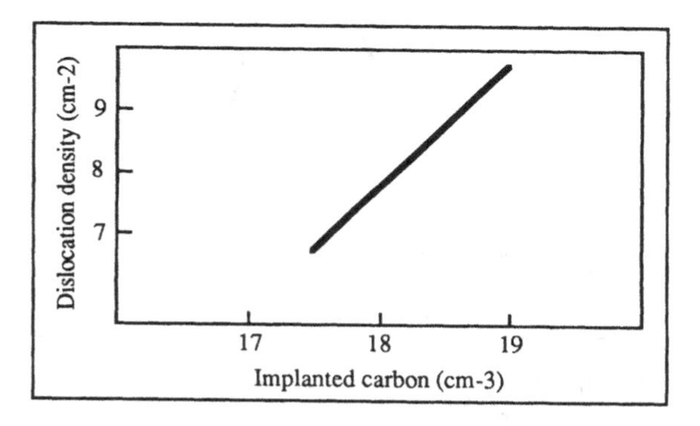

Figure 3.3.1: Dislocation density in SIMOX annealed at 1300°C as a function of implanted carbon concentration.

3.3.2. Oxygen contamination

The silicon overlayer of SIMOX material annealed at relatively low temperature (1150°C) contains large amounts of oxygen. Even moderate concentrations of oxygen (10^{17} cm^{-3}) can induce undesirable modifications of the silicon resistivity. Indeed, electrons are generated upon annealing of oxygen-containing silicon. These carriers are commonly referred to as thermal donors or new donors, depending on the temperature range in which they are activated (around 450°C for thermal donors and 750°C for new donors). It is worth noting that 400-450°C is the usual temperature used for metal sintering, which is the last thermal operation in an integrated circuit fabrication process. Consequently, the thermal donors generated during this step will be fully activated in the finished product. The generation rate of thermal donors in 1150°C-annealed SIMOX films is 10^{11} cm^{-3} s^{-1}, and it is 10^{12}

$cm^{-3}\ s^{-1}$ for new donors. Thermal donor formation is explained by the formation of oxygen complexes, while the origin of new donors remains unclear but may be related to the formation of surface states between the silicon overlayer matrix and carbon-oxygen complex precipitates [3.29]. Figure 3.3.2 presents the resistivity of 1150°C-annealed SIMOX material as a function of subsequent annealing temperature. The annealing time was 30 minutes in all cases, and the sheet resistivity was measured by four-point probe and spreading resistance techniques [3.29].

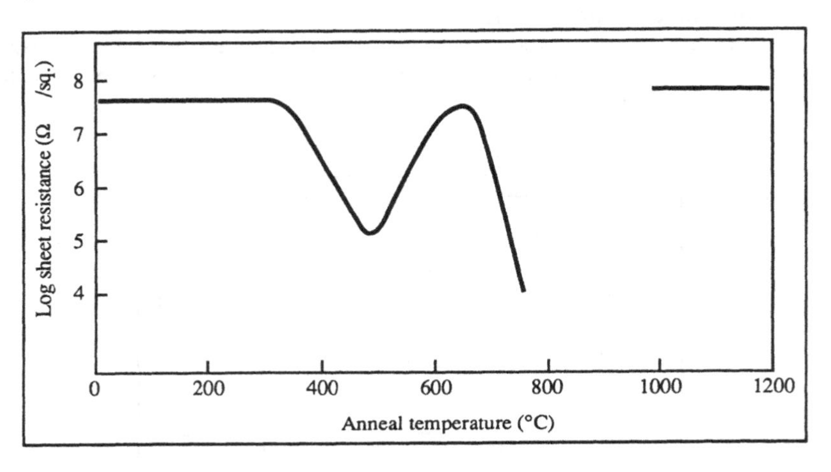

Figure 3.3.2: Sheet resistance of the silicon overlayer as a function of annealing temperature.

3.4. Carrier lifetime and surface recombination

The lifetime of minority carriers is a measure of the quality of the silicon films. The lifetime is affected by the presence of both crystal defects and metallic impurities. There exists no non-destructive method as such for measuring the lifetime in the silicon overlayer of an SOI structure. The measurement techniques rely either on the measurement of the lifetime in devices such as MOSFETs realized in the SOI material or on the measurement of the lifetime in the underlying silicon substrate and on the correlation between the lifetime in the substrate and the level of metal contamination within the silicon overlayer.

3.4.1. Surface photovoltage

The surface photovoltage (SPV) technique relies on the generation of a voltage at the surface of a silicon sample upon illumination. SPV uses a chopped beam of monochromatic light of photon energy hν slightly larger than the band gap E_G of silicon. Electron-hole pairs are produced by the absorbed photons. Some of these pairs diffuse to the illuminated surface where they are separated by the electric field of the surface space-charge region whose thickness is *w*, thereby producing a surface photovoltage ΔV. A portion of ΔV is capacitively coupled to a transparent conducting electrode adjacent to the illuminated surface (Figure 3.4.1). This signal is then amplified to produce a quasi-dc analog output which is proportional to ΔV.

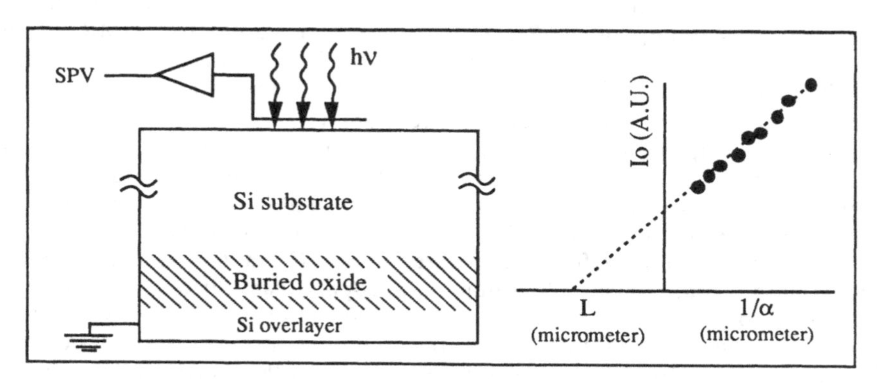

Figure 3.4.1: Experimental set-up for SPV measurement of an as-implanted SIMOX wafer (left) and graphical method for finding the carrier diffusion length (right).

The value of ΔV is a function of the excess minority carrier (holes in the case of n-type silicon) density Δp(0) at the edge of the surface space-charge region. This density Δp(0) is in turn dependent on the incident light flux I_o, the optical absorption coefficient α, the optical reflectance at the illuminated surface ρ, the recombination velocity at the illuminated surface *S*, as well as the diffusion length *L*. A steady-state solution of the one-dimensional diffusion equation for the excess carrier density is given by $\Delta p(0) = \frac{I_o(1-\rho)}{D/L + s}\ \frac{\alpha L}{1+\alpha L}$ [3.30] if one assumes $\alpha w \ll 1$, $w \ll L$ and $\Delta p \ll n_o$, where n_o is the majority carrier density. D is the minority carrier diffusion coefficient. One also assumes that $\alpha W \gg 1$, where W is the thickness of the silicon wafer. A series of different wavelength values is selected to yield different values of α. At each wavelength, I_o is adjusted to give the same value (constant magnitude) of ΔV. As a consequence, the value of Δp(0) is constant as well. If ρ is essentially constant over the wavelength region of interest, the above equation may be rewritten: $I_o = C[1+(\alpha L)^{-1}]$, where *C* is a constant. If I_o is plotted against α^{-1} for each constant-magnitude ΔV point, the result is a linear graph whose extrapolated intercept on the negative α^{-1} axis is L (Figure 3.4.1) [3.31]. The carrier lifetime, τ_p, can be deduced from the relationship $L^2=D\tau$, with $D = \frac{kT}{q}\mu$, where μ is the carrier mobility.

The SPV technique cannot be directly used to measure the lifetime in the silicon overlayer of an SOI sample because the silicon film is too thin for the condition $\alpha W>>1$ to be met. SPV can nevertheless be employed as an indirect measurement of the heavy metal contamination of SOI material. An excellent correlation is observed between the metallic impurity concentration (Fe, Cr, Ni and Cu) in the silicon overlayer of as-implanted SIMOX wafers, measured by SIMS or by spark source mass spectrometry, and the SPV diffusion length measured in the silicon substrate [3.32]. The actual relationship between the lifetime L and the heavy metal concentration C is in the form $L \cong 1/C$. The relationship between the metal concentration and the lifetime in the substrate can be explained as follows. Metal impurities are introduced in the wafer during the oxygen implantation process. In the experiment reported in ref. [3.32], the implantation took 6 hours, and was carried out at a wafer temperature of 580°C. In 6 hours, and at this temperature, iron, nickel, chromium and copper can diffuse in silicon at distances of 650, 3000, 200 and 6500 micrometers, respectively. Therefore, one can assume that the metal concentration is constant throughout the depth of the substrate, and that the concentration in the silicon film is correlated to that in the substrate. SPV is used to routinely monitor the heavy metal impurity level in SIMOX wafers [3.33].

3.4.2. Lifetime measurement in devices

The minority carrier lifetime is an important parameter which has significant impact on device characteristics. Indeed, junction leakage appears if the generation lifetime is short (see *e.g.* Section 7.2.1). A high recombination lifetime, on the other hand, enhances parasitic bipolar problems (see Section 5.7). Minority carrier lifetime is affected by device processing, and its value after device fabrication may be significantly different from its value before processing.

3.4.2.1. Generation lifetime and surface generation

The minority carrier generation lifetime can be extracted from leakage current measurements [3.34, 3.35] and Zerbst-like techniques [3.36, 3.37]. A first method is based on the combined measurement of the leakage current and the capacitance of a junction as a function of the diode reverse bias. This method was demonstrated in thick SOI with non-reach-through junctions [3.34] (Figure 3.4.2).

The current flowing through a reverse-biased (P^+-N) junction consists of several components, which include the diffusion current, I_{diff}, the bulk generation current I^b_{gen}, the surface generation current I^s_{gen}, and the field-enhanced generation current, I_{fe}. The last term involves field-dependent current not included in the other three current components.

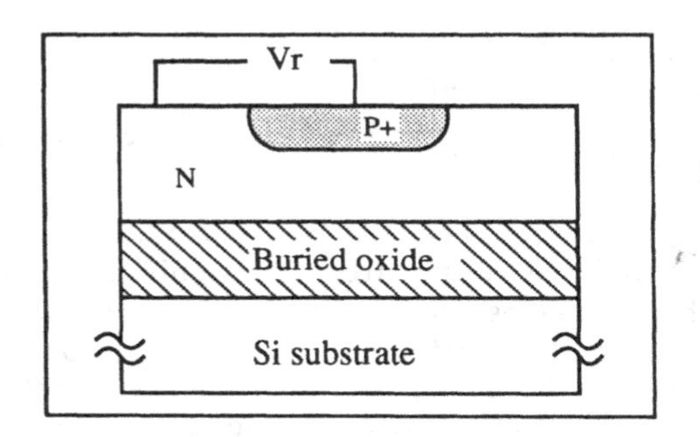

Figure 3.4.2: P^+-N diode for lifetime measurement in thick SOI film.

If the applied reverse bias V_r is such that the depletion region underneath the junction does not reach through the buried oxide, the total reverse current, I_r, is given by:

$$I_r(V_r) = A_s q n_i S_o + A_b q \int_0^{W_i(V_r)} \frac{n_i}{\tau_g(x)} dx + I_{diff}(V_r) + I_{fe}(V_r) \quad (3.4.1)$$

where S_o is the surface generation velocity, A_b is the area of the space-charge region in the bulk of the silicon, A_s is the area of the diode junction, n_i is the intrinsic carrier concentration, and W_i is the depletion depth in which generation takes place. W_i remains smaller than the depletion depth in the N-type silicon film, W, up to voltages so high that I_{fe} is no longer negligible. It can be expressed by [3.34]:

$$W_i = \sqrt{\frac{2\varepsilon_{si} kT}{q^2 N_d}} \left(\sqrt{\ln\frac{N_d}{n_i} + \frac{qV_r}{kT}} - \sqrt{\ln\frac{N_d}{n_i}} \right) \quad (3.4.2)$$

where N_d is the n-type dopant concentration in the silicon film. If V_r is not too large, I_{diff} and I_{fe} are negligible. I^s_{gen} is independent of V_r.

Using $W=\sqrt{\frac{2\varepsilon_{si}}{qN_d}(\phi+V_r)}$, $C=A_b\varepsilon_{si}/W$ and thus $dW/dV_r=(\varepsilon_{si}A_b/C^2).dC/dV_r$, where C is the junction capacitance, W is the depletion depth, and ϕ is the junction built-in potential, equation (3.4.1) can be expressed in a differential form:

$$\tau_g(W_i)=\frac{\varepsilon_{si}qn_iA_b^2}{C^2}\frac{dW_i}{dW}\frac{dC/dV_r}{dI_r/dV_r} \qquad (3.4.3)$$

where $\frac{dW_i}{dW}=\left(\frac{V_r+\frac{kT}{q}\ln\frac{N_aN_d}{n_i^2}}{V_r+\frac{kT}{q}\ln\frac{N_d}{n_i}}\right)^{1/2}$ is a correction factor which takes into account the difference between W_i and W. It is worthwhile noting that even though W_i may differ from W significantly, $dW_i/dW\cong1$ for a wide range of bias voltages and doping concentrations. N_a is the doping concentration of the P^+ diffusion. The generation lifetime can be obtained from (3.4.3) through performing measurement of both reverse leakage current and junction capacitance and can be plotted as a function of depth in the silicon film. It is experimentally observed that the measured lifetime drops significantly within a Debye length ($\cong$100 nm) of the silicon-buried oxide interface, due to the contribution of the surface generation. Study of the surface generation velocity can be performed by performing the measurement for different values of back-gate voltage. If the back interface is either in accumulation or inversion, only those generation centers which are within the silicon film contribute to the generation current. If, on the other hand, the back interface is depleted, generation at the interface provides yet another contribution to the total generation current, which results in a peak in $I_r(V_{G2})$ characteristics (Figure 3.4.3) [3.34-3.35].

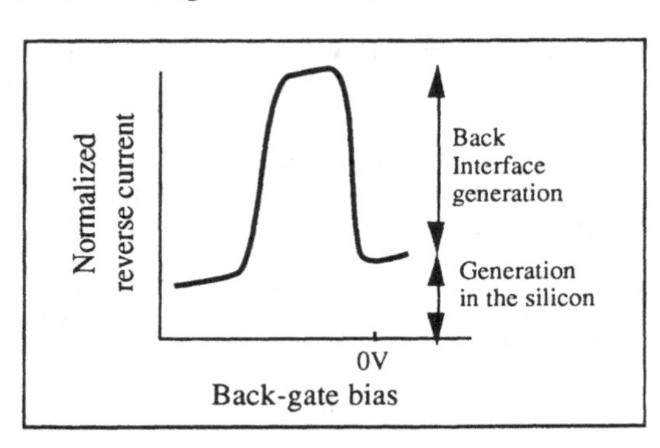

Figure 3.4.3: Reverse diode current as a function of back-gate bias.

Another method is more appropriate for the measurement of the minority carrier generation lifetime in thin silicon films. It is based on the measurement of the time required to create an inversion layer in a deep-depletion (or accumulation-mode) MOSFET. The minority carriers needed to create the inversion layer can only come from generation in the bulk of the silicon film or at the $Si\text{-}SiO_2$ interfaces. The dc characteristics of such a device are described in Section 5.8. We will consider the case of an n-channel ($N^+\text{-}N^-\text{-}N^+$) device [3.36, 3.37]. From equation (5.8.7), and taking into account both the presence of interfaces traps and the formation of an inversion layer (for large negative gate biases), one finds:

$$V_G - V_{FB} = \frac{qN_d x_{depl}^2}{2\varepsilon_{Si}} + \frac{qN_d x_{depl}}{C_{ox}} + \frac{Q_i + Q_{it}}{C_{ox}} \qquad (3.4.4)$$

where Q_i is the inversion charge and Q_{it} is the surface-state charge. N_d is the n-type dopant density in the channel region.
One can also write:

$$x_{depl}(t) = t_{si} - x_a = t_{si} - \frac{L}{\sigma\, W\, V_{DS}} I_D(t) = t_{si} - B\, I_D(t) \qquad (3.4.5)$$

where $B = \frac{L}{\sigma\, W\, V_{DS}}$ if the device is operating in the linear regime (V_{DS} is small), and where x_a is the width of the non-depleted portion of the silicon film, and σ is the conductivity of the N^- silicon [3.37]. L is the gate length, and W is the device width. Equation (3.4.4) can be rewritten:

$$Q_i(t) + Q_{it}(t) = C_{ox}\,(V_G - V_{FB}) - \frac{qN_d C_{ox}}{2\varepsilon_{Si}}\,(x_{depl}(t))^2 - qN_d x_{depl}(t)$$

or

$$Q_i(t) + Q_{it}(t) = C_{ox}\,(V_G - V_{FB}) - \frac{qN_d C_{ox}}{2\varepsilon_{Si}}\left(\left(x_{depl}(t) + \frac{\varepsilon_{Si}}{C_{ox}}\right)^2 - \left(\frac{\varepsilon_{Si}}{C_{ox}}\right)^2\right) \qquad (3.4.6)$$

Using the classical model for generation of carriers in a semiconductor and the definition of surface generation velocity, we can write [3.38]:

$$\frac{d(Q_i(t) + Q_{it}(t))}{dt} = \frac{q\, n_i}{\tau_{gen}}\Big(x_{depl}(t) - x_{depl}(t{=}\infty)\Big) + q\, n_i\, S_o \qquad (3.4.7)$$

where τ_{gen} is the generation lifetime in the silicon film, and S_o is the surface generation velocity. Derivating (3.4.6) and combining with (3.4.7) and (3.4.5), one obtains:

$$F(t) = \frac{B}{\tau_{gen}}\Big(I_D(t{=}\infty) - I_D(t)\Big) + S_o \qquad (3.4.8)$$

where

$$F(t) = - \frac{N_d C_{ox}}{2 n_i \varepsilon_{Si}} \frac{d}{dt}\left(\left(t_{Si} + \frac{\varepsilon_{Si}}{C_{ox}}\right) - B I_D(t)\right)^2 \qquad (3.4.9)$$

Plotting F(t) as a function of $\left(I_D(t=\infty) - I_D(t)\right)$ yields the values of both τ_{gen} (slope of the dotted line) and S_o (y-axis intercept) (Figure 3.4.4.B).

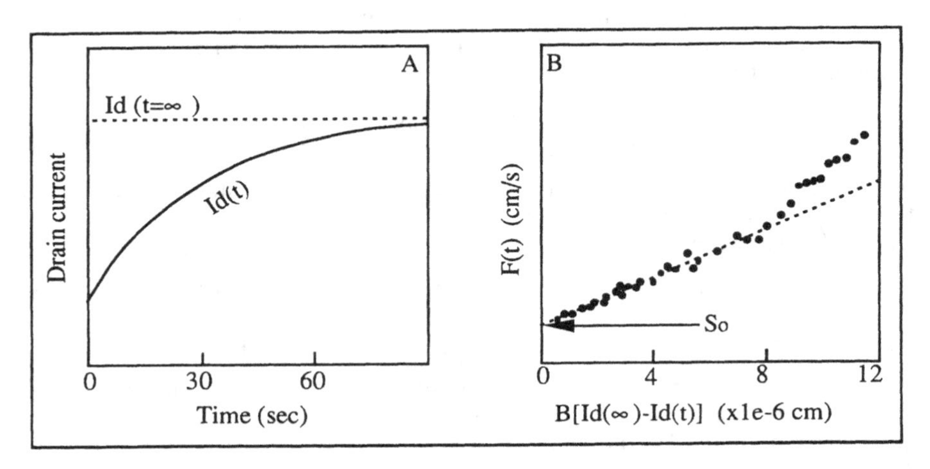

Figure 3.4.4: Drain current as a function of time in a depletion SOI MOSFET pulsed into inversion (A) and extraction of τ_{gen} and S_o from the slope and the intercept of equation (3.4.9) (B).

This model assumes that the surface generation velocity is constant, which is true only when the silicon interface underneath the gate is in inversion, which condition yields low values of S_o. A better value of the surface generation is obtained by taking S_o equal to the difference between the measured data and the extrapolated line of Figure 3.4.4.B. A graph of $S_o(t)$ can then be produced Figure (3.4.5.A).

The surface generation velocity is highest right after the application of the gate bias (no inversion layer is present to act as a screen between the interface and the depletion layer). Its value then decreases with time, as the inversion layer forms (Figure 3.4.5.A). Similarly, the effective generation lifetime can be plotted as a function of the depth in the silicon film. Indeed, the depth of the depletion zones varies with the magnitude of the applied gate bias. Small gate biases allow one to measure the effective lifetime near the surface (τ_{gen} is low because of the presence of the front interface - Figure 3.4.5.B), larger biases measure it deeper in the silicon film (τ_{gen} increases), and yet larger biases can be used to probe the back interface (τ_{gen} decreases again).

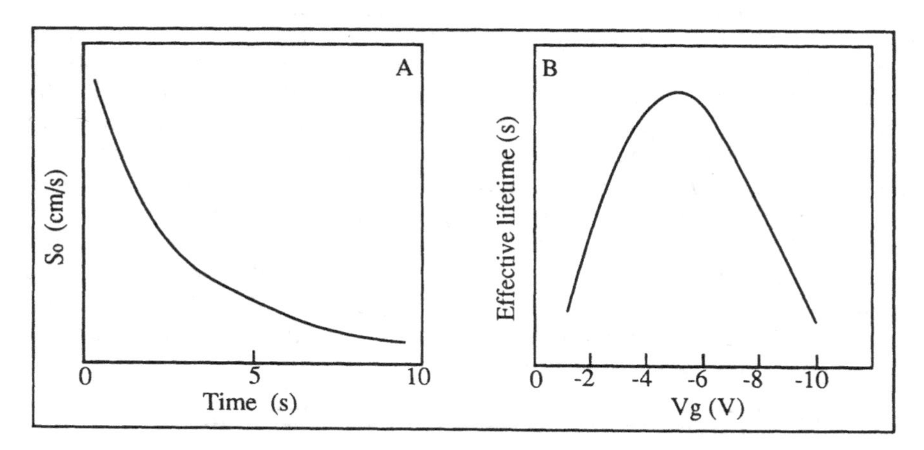

Figure 3.4.5: Surface generation as a function of time after gate biasing (A) and effective generation lifetime as a function of gate bias (B).

The above relationships were developed for edgeless devices. If a device with edges is measured, generation at the edges has to be taken into account, and equation (3.4.8) must be written [3.39, 3.40]:

$$F(t) = \frac{B}{\tau_{gen}}\big(I_D(t{=}\infty) - I_D(t)\big) + \frac{2\ L\ S_{edge}}{L\ \ W} B \big(I_D(t{=}\infty) - I_D(t)\big) + S_o$$

or

$$F(t) = \frac{B}{\tau_{eff}}\big(I_D(t{=}\infty) - I_D(t)\big) + S_o \qquad (3.4.10)$$

such that one obtains:

$$\frac{1}{\tau_{eff}} = \frac{1}{\tau_{gen}} + \frac{2S_{edge}}{W} \qquad (3.4.11)$$

where W is the width of the device and S_{edge} is the surface generation velocity at the edges of the device. Typical generation lifetime values are 1-10 μs in good-quality SIMOX material and 40 μs in subgrain-boundary-free ZMR material [3.37, 3.36].

3.4.2.2. Recombination lifetime

The recombination lifetime, or more exactly the effective recombination lifetime (which includes the influence of interface recombination), is a critical parameter for all characteristics involving parasitic bipolar effects. Its value can be estimated through the measurement of the gain β of lateral bipolar transistors with different base widths. Indeed, based on the bipolar transistor theory, one can write (equation 5.7.2): $\beta \equiv 2\ (L_n/L_B)^2 - 1$, where L_B is the base width, which can be assumed, in first approximation, to be equal to the effective channel length, L_{eff}, and L_n is the electron diffusion length (we consider here the case of a NPN (n-channel) device). From the relationship $L_n{}^2 = D_n\ \tau_n$, where D_n and τ_n are the diffusion coefficient and the effective recombination lifetime of the minority carriers (electrons) in the base, respectively.

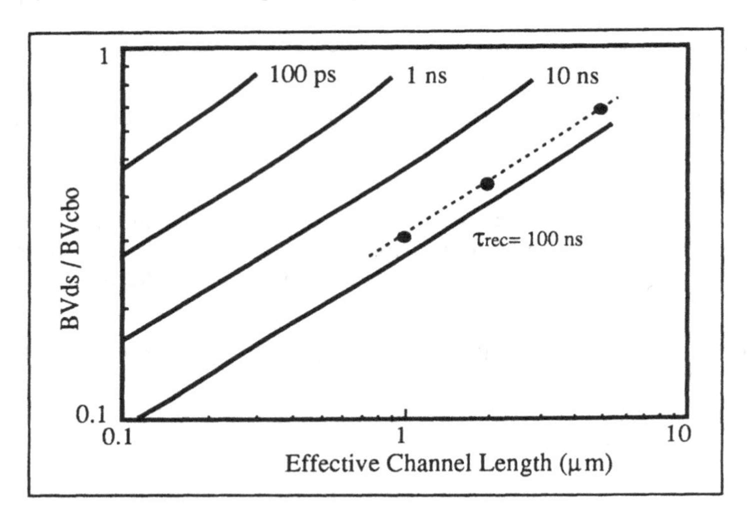

Figure 3.4.6: Determination of the effective recombination lifetime by means of drain breakdown voltage measurements. The circles represent experimental data.

If lateral bipolar transistor structures (= MOSFETs with body contacts) are not available, an estimation of the effective recombination lifetime can be obtained from a graph representing the drain breakdown voltage of n-channel SOI MOSFETs as a function of the effective gate length [3.41]. Indeed, experimental values of BV_{DS}/BV_{CBO}, can be plotted as a function of channel length on an abacus such as that which is presented in Figure 5.7.3 of Section 5.7. BV_{DS} is the measured drain breakdown voltage, and BV_{CBO} is the intrinsic breakdown voltage of the drain junction, which is approximately equal to the BV_{DS} of a very long channel device ($L_{eff} \cong L$ = 50...100 μm). Figure 3.4.6 illustrates the method and gives the example of measurements carried out on devices with an effective recombination lifetime τ_{rec} of approximately 70 nsec.

3.5. Silicon-oxide interfaces

3.5.1. Capacitance measurements

Classical C-V techniques can be employed to measure the charges in the oxide layers and at the Si-SiO_2 interfaces of SOI devices, but the interpretation of the data is rather difficult. Indeed, the direct measurement of the capacitance of the whole SOI structure yields a complicated C-V curve (Figure 3.5.1) in which one can find contributions from the accumulation, depletion and inversion states at the front and back interfaces of the silicon film as well as at the top of the silicon substrate [3.42]. Although analytical models of the SOI capacitor are available [3.42], it is usually easier to simulate the metal-oxide-silicon-oxide-silicon (MOSOS) structure numerically and to compare the simulation results with the measurements in order to obtain an estimation of the charges in the oxide and at the Si/SiO_2 interfaces [3.43].

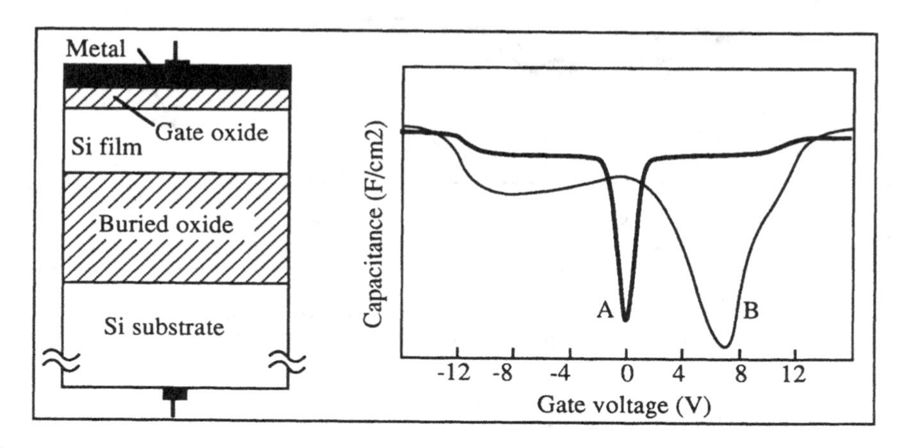

Figure 3.5.1: The SOI structure and an example of quasistatic C-V curves measured between the top metal electrode and the substrate. A: curve with one minimum [3.43] ; B: curve with two minima [3.42].

It is possible to measure separately the different capacitance components from MOS capacitors fabricated using a conventional SOI-CMOS process. Interdigitated capacitors are often used to minimize the parasitic resistance of the lightly doped channel region. Figure 3.5.2 presents the schematic configuration of such a capacitor. One assumes that the silicon film underneath the gate is not fully depleted. C_1 is the gate capacitance, C_2 is the capacitance across the buried oxide underneath the gate region, C_3 is the capacitance across the buried oxide underneath the P+ diffusion, and C_4 is the capacitance between the metal patterns (line + pad) and the substrate (Figure 3.5.2). The capacitance C_4 is actually composed of two capacitances, C_{4F} and C_{4G} corresponding to the metal lines needed to contact the P+ diffusion (film contact) and the gate, respectively. The equivalent circuit representing the structure can be reduced to three capacitors, C_A, C_B and C_C, where

$$\begin{aligned} C_A &= C_1 \\ C_B &= C_2 + C_3 + C_{4F} \\ C_C &= C_{4G} \end{aligned} \tag{3.5.1}$$

Access to the capacitors is obtained through three terminals: the gate pad, the film contact pad, and the silicon substrate (Figure 3.5.2) [3.44].

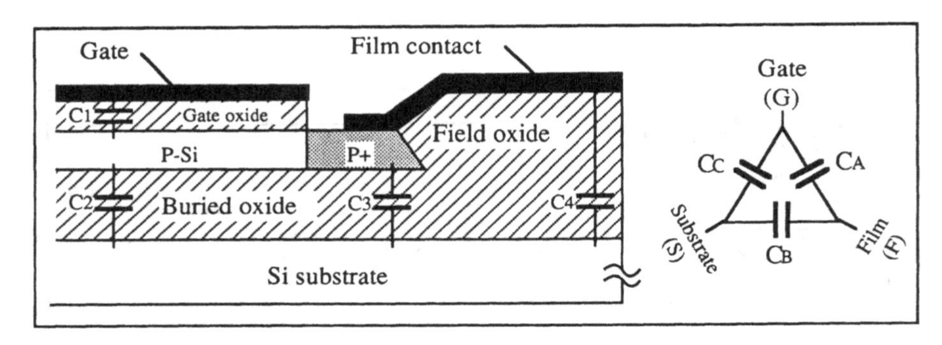

Figure 3.5.2: Schematic configuration of an SOI capacitor fabricated using a standard CMOS process and equivalent capacitor network.

If capacitance measurement is carried out between two of these terminals and leaving the third one floating, (*e.g.* measuring C_1 between the gate and film electrodes), altered C-V curves are obtained due to the presence of the two other capacitors. Better results are obtained by measuring C-V curves between one contact and the two others sorted to one another. For instance, $C_{G/FS}$ is measured between the gate terminal and the substrate and film connected together. From the equivalent circuit of Figure 3.5.2, we have for the three possible measurement configurations:

$$C_{G/FS} = C_C(V_{GS}) + C_A(V_{GF})$$
$$C_{F/SG} = C_A(-V_{GF}) - C_B(V_{FS})$$
$$C_{S/GF} = C_B(-V_{FS}) + C_C(-V_{GS}) \quad (3.5.2)$$

Combining these equations, one obtains:

$$C_A(V_{GF})+C_A(-V_{GF}) = C_{G/FS}(V_{G/FS})+C_{F/SG}(V_{F/SG})-C_{S/GF}(-V_{S/GF}) \quad (3.5.3)$$

as well as two similar expressions for C_B and C_C, obtained by circular permutations. Unfortunately, C_A, C_B and C_C cannot be derived independently from the above relationships because they are asymmetrical with respect to the applied bias voltage (i.e., $C_A(V_{GF}) \neq C_A(-V_{GF})$). Nevertheless, $C_C=C_{4G}$ can be measured independently using a metal field capacitor, such as an unconnected metal pad, of capacitance C_T. We then obtain: $C_C(V_{GS})=C_{4G}=C_T(V_{GS})\frac{A_G}{A_T}$ and $C_{4F}=C_T\frac{A_F}{A_T}$, where A_G, A_F, and A_T are the areas of the capacitors C_{4G}, C_{4F}, and C_T, respectively. Now, C_A and C_B can easily be found using (3.5.2), and one obtains from (3.5.1): $C_1=C_A$ and $C_2+C_3 = C_B-C_{4F}$ [3.44]. Separation of C2 and C3 can be obtained by measuring two test structures with different area ratios for the capacitors C_2 and C_3. Finally, C-V analysis can be performed on the C_1 and C_2 curves.

3.5.2. Charge pumping

The charge pumping technique [3.45, 3.46] is very efficient for characterizing Si/SiO_2 interfaces. It can be used with small area devices and can yield the distribution of interface states in the band gap. In the case of SOI devices, the front and back interfaces can be characterized independently from one another. The principle of the charge pumping technique in an bulk device is the following. Source and drain are connected together and slightly reverse biased, with respect to the substrate. A periodical triangular or trapezoidal signal, ΔV_G, with frequency *f* is applied to the gate. ΔV_G is sufficiently large to switch the silicon surface underneath the gate from accumulation to strong inversion (Figure 3.5.3). When the device goes into inversion, minority carriers are provided by the source and the drain to form the inversion channel. Some of these carriers are trapped by interface states. When the gate voltage is switched to produce an accumulation layer in the device, the inversion carriers (electrons in an n-channel device) disappear swiftly towards source and drain, and the minority carriers (electrons) which were trapped in the surface states now recombine with majority carriers from the substrate. This hole current constitutes the charge-pumping current I_{cp}.

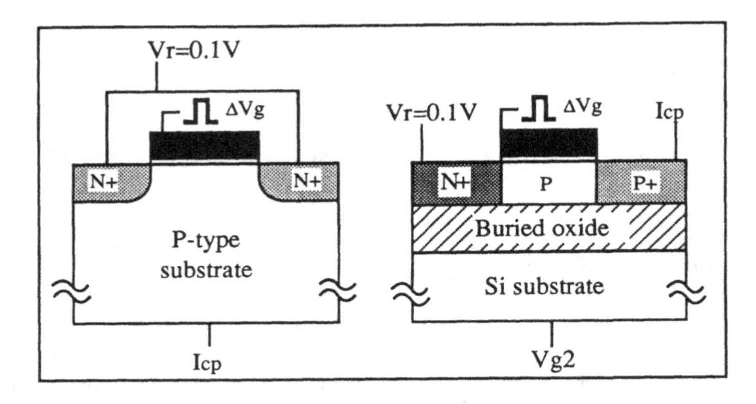

Figure 3.5.3: Experimental set-up for charge-pumping measurement in a bulk MOSFET (left) and an SOI gated PIN diode (right). I_{cp} is the charge-pumping current.

If the gate is pulsed at a frequency *f*, the charge-pumping current is given by:

$$I_{cp} = q^2 \, \overline{N_{it}} \, A \, \Delta\Phi_s \, f$$

where $\overline{N_{it}}$ is the average interface trap density, A = W.L is the channel area, and $q\Delta\Phi_s$ is the energy range scanned within the bandgap. This basic expression can be extended to more sophisticated pumping techniques using gate offset bias and trapezoidal gate pulsed with different rise and fall times, in which case the energy distribution of the surface states across the bandgap can be obtained [3.46].

In the case of an SOI MOSFET, the charge-pumping current can be measured through a substrate contact. Better results are, however, obtained by carrying the measurement on a PIN (or $P^+P^-N^+$) gated diode (Figure 3.5.3) [3.47]. In inversion, minority carriers are provided by the N^+ cathode, and the charge-pumping current is measured at the P^+ anode. Measurement of the front-interface trap density is obtained by pulsing the front gate, while back-interface trap density can be measured by pulsing the silicon substrate (back gate). Figure 3.5.4 presents the charge-pumping current measured on a SIMOX PIN gated diode as a function of the pulse frequency. Front and back interfaces were measured separately, by pulsing either the front gate or the back gate. The charge pumping current is larger when the back interface is probed, indicating a larger density of traps, $\overline{N_{it}}$, than at the front interface. One can also observe that the back interface states is composed of two components: a high density of slow states having a cutoff frequency of ≅1-10 kHz and a lower density of fast states, indicated by the dotted line in Figure 3.5.4 [3.47].

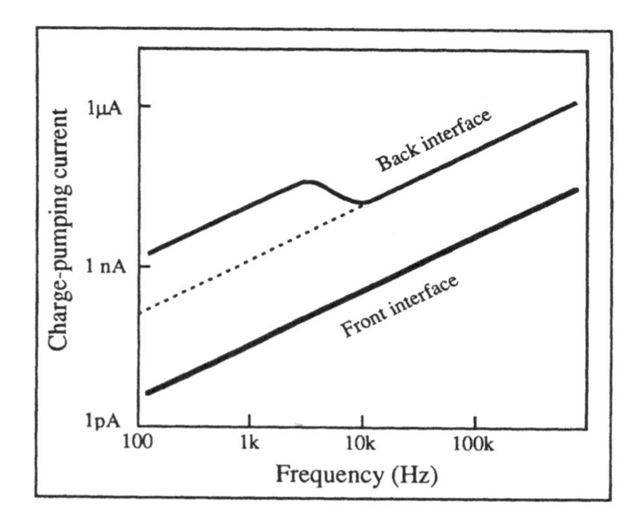

Figure 3.5.4: Charge-pumping current measured on a gated SIMOX PIN diode, as a function of frequency.

CHAPTER 4 - SOI CMOS Technology

Complementary MOS (CMOS) is, by far, the technology of choice for the realization of integrated circuits on SOI substrates. This Chapter will compare CMOS processing on bulk silicon and on SOI wafers. Processing of both thin and thicker SOI films will be discussed. We will assume that circuit processing is carried out on commercially available substrates, such as SIMOX wafers. It is worthwhile keeping in mind that, unlike in the case of SOS, SOI wafers contain only silicon and silicon dioxide, and that the appearance of SOI wafers is very similar to that of bulk silicon wafers. As a consequence, SOI circuit processing can be carried out in standard bulk silicon processing lines. Mixed batches (containing both bulk and SOI substrates) can be processed as well.

4.1. Comparison between bulk and SOI processing

Processing techniques for the fabrication of CMOS circuits in bulk silicon and in SOI are very similar. Figure 4.1.1 presents cross-sections of CMOS inverters made in bulk (p-well technology), in "thick-film" SOI (200-500 nm-thick films), and in thin-film SOI.

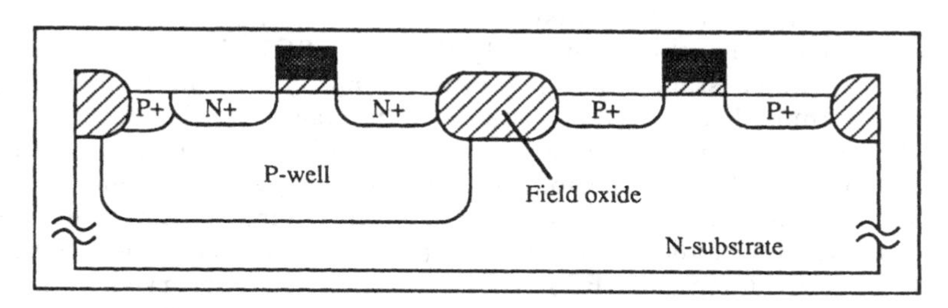

Figure 4.1.1.A: Cross-section of a bulk CMOS inverter.

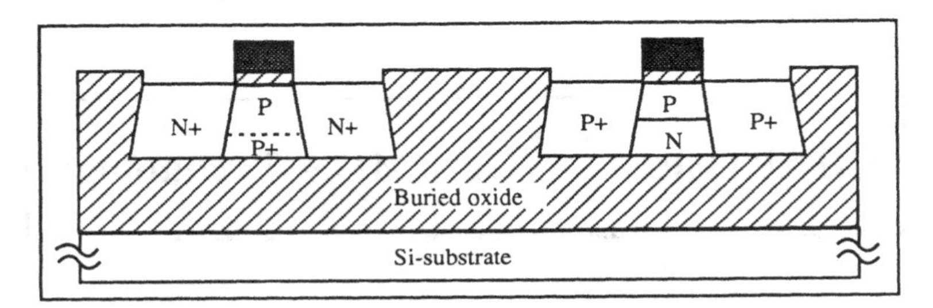

Figure 4.1.1.B: Cross-section of a "thick-film" SOI CMOS inverter.

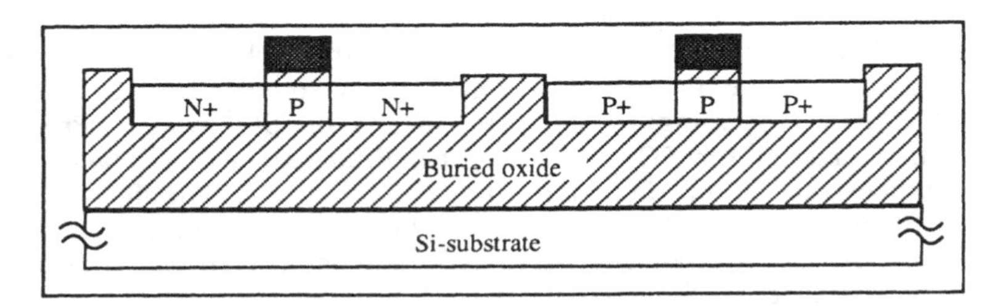

Figure 4.1.1.C: Cross-section of a thin-film SOI CMOS inverter.

The cross sections of Figure 4.1.1 are, of course, quite schematic. For instance, bulk CMOS can be much more complicated than in Figure 4.1.1.A and can make use of an epitaxial substrate, twin wells, or retrograde wells. From the cross sections, it is obvious that SOI processing, and more specifically thin-film SOI processing, is simpler than bulk processing. For instance, there is no need to create diffused wells in SOI. The anti-punchthrough implant used in bulk CMOS is kept unchanged in the case of a p-channel SOI device, and is replaced by a back-channel leakage suppression implant for the n-channel device (Figure 4.1.1.B). In thin-film, fully-depleted SOI devices, these deep implants are unnecessary, and the entire impurity profile in the channel area is determined by a single shallow implant. In the frequent case where N^+ polysilicon is used as gate material, p-type impurities (boron) are used to control the threshold voltage in both the n-channel and the p-channel devices (Figure 4.1.1.C). Table 4.1.1 compares simplified CMOS process flows for bulk, "thick-film", partially-depleted (PD) SOI, and thin-film, fully-depleted (FD) SOI. It can be observed that SOI processing is simpler (less steps), and that at least one mask step is saved (two mask steps can be saved if the p-channel anti-punchthrough implant is a blanket implant, later compensated in the n-channel devices by a "heavy" back channel-stop implant). The simplification of the process is even more dramatic if thin-film (100 nm or less) devices are fabricated.

Bulk CMOS	PD SOI CMOS	FD SOI CMOS
Oxidation	Oxidation	Oxidation
Well litho*		
Well doping & drive-in		
Nitride deposition	Nitride deposition	Nitride deposition
Active area litho*	Active area litho*	Active area litho*
Nitride etch	Nitride etch	Nitride etch
Field implant litho*	Field implant litho*	Field implant litho*#
Field implant	Field implant	Field implant#
Field oxide growth	Field oxide growth	Field oxide growth
Nitride strip	Nitride strip	Nitride strip
P-channel litho*	P-channel litho*	
Anti-punchthru implant	Anti-punchthru implant	
Gate oxide growth	Gate oxide growth	Gate oxide growth
P-ch Vth implant	P-ch Vth implant	P-ch Vth implant
N-channel Vth litho*	N-channel Vth litho*	N-channel Vth litho*
Anti punchthru implant	Back channel implant	
N-ch Vth implant	N-ch Vth implant	N-ch Vth implant
Poly deposition & doping	Poly deposition & doping	Poly deposition & doping
Gate litho* & etch	Gate litho* & etch	Gate litho* & etch
P+ S&D litho*	P+ S&D litho*	P+ S&D litho*
P+ S&D implant	P+ S&D implant	P+ S&D implant
N+ S&D litho*	N+ S&D litho*	N+ S&D litho*
N+ S&D implant	N+ S&D implant	N+ S&D implant
S&D reoxidation	S&D reoxidation	S&D reoxidation
Dielectric deposition	Dielectric deposition	Dielectric deposition
Contact hole litho*	Contact hole litho*	Contact hole litho*
Contact hole opening	Contact hole opening	Contact hole opening
Metallization	Metallization	Metallization
Metal litho*	Metal litho*	Metal litho*
Metal patterning	Metal patterning	Metal patterning
Sintering	Sintering	Sintering

Table 4.1.1: Comparison of bulk, "thick-film" (partially depleted - PD) SOI and thin-film, fully-depleted (FD) CMOS process flows. N^+ polysilicon is used as gate material. (The "*" symbol indicates a lithography step, the "#" indicates that the step may be optional.)

We will now describe into more detail some particularities of SOI processing: isolation techniques, doping profiles, and S&D silicidation.

4.2. Isolation techniques

There are many different ways of isolating the active silicon islands from one another. All these are simpler than the isolation schemes used in bulk silicon technology, owing to the presence of a buried insulator underneath the silicon film which provides an intrinsic vertical isolation. In a sense, one could say that a part of the complexity of the device isolation process used in bulk technology has been transferred to the wafer manufacturing stage (the fabrication of silicon-on-insulator material). Three of the main isolation techniques will be described next.

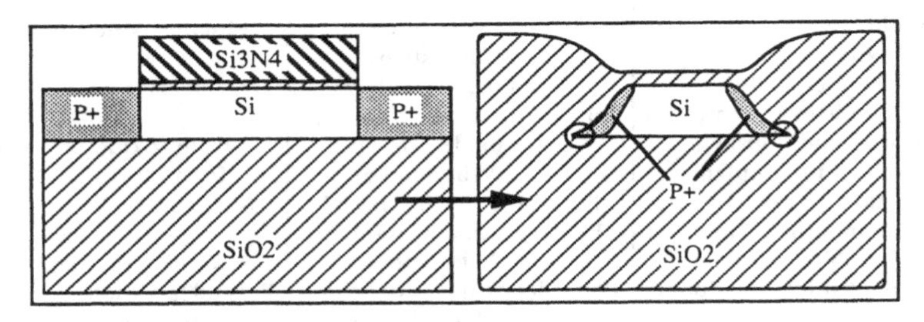

Figure 4.2.1: LOCOS isolation. The circles at the edges of the silicon island (left) indicate where source-to-drain edge leakage may occur.

The LOCOS (local isolation of silicon) technique is undoubtedly the most popular isolation scheme used in bulk CMOS. It is a well-known, reliable and well-controlled process presenting little yield hazards. It consists in the following steps: a thin "pad" oxide is grown, and a silicon nitride layer is deposited. Using a mask step, the nitride is patterned to define the active silicon areas. Boron is then implanted around those islands which will contain n-channel devices (channel-stop implant), and the field oxide is thermally grown. No oxide is grown where the silicon is protected by nitride. A nitride and pad oxide stripping step completes the process. The exact same process can be used in SOI for lateral device isolation. During the field oxide growth step, the silicon film which is not covered with nitride is completely consumed, and the thermally-grown field oxide reaches through to the buried oxide (Figure 4.2.1).

The lateral encroachment of the field oxide, called "bird's beak", is proportional to the thickness of the grown oxide. This encroachment has typical values of a fraction of a micrometer, and hampers the use of the LOCOS technique for submicron bulk processing, where 0.5...0.8 μm-thick field oxides are grown. The local stresses in the silicon is also related to the thickness of the field oxide. In an SOI process the thickness of the oxide which must be grown to fully isolate the silicon islands is 2.5 - 3 times the thickness of the silicon film. If a 100 nm-thick SOI film is used, one has to grow a field oxide of only 250-300 nm. This implies less stress in the silicon, and, most importantly, a significantly smaller bird's beak. As a consequence, it seems that the LOCOS isolation can be used to realize devices with smaller geometries in SOI than in bulk.

In bulk devices, the P^+ implant (channel-stop implant) which is carried out prior to growing the field oxide prevents surface inversion of the silicon underneath the field oxide. Such an inversion would lead to leakage between n-channel devices. In SOI devices, the same implant is used for preventing source-to-drain leakage from taking place along the edges of the silicon island. Indeed, an inversion layer can form at the tip of the bottom corners of the islands (Figure 4.2.1). This inversion layer degrades the subthreshold slope of the device and is the source of leakage current when the device is turned off.

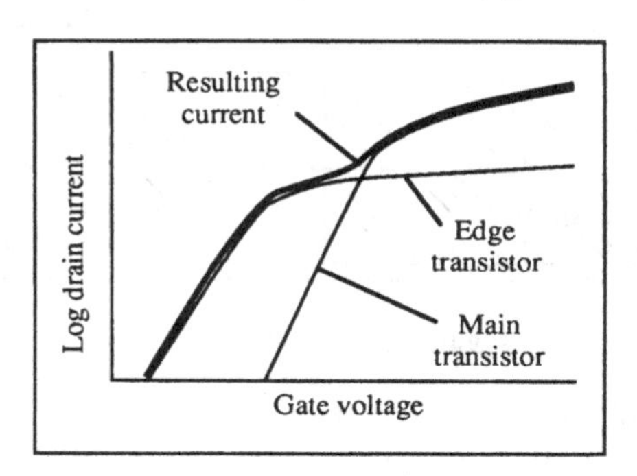

Figure 4.2.2: Subthreshold characteristics of a transistor with edge leakage.

The edge leakage current is caused by the presence of a parasitic edge transistor placed in parallel with the main transistor. It has the same gate length as the main device, but has a much smaller width. If the P+ impurity concentration at the edges of the silicon island is high enough, the threshold voltage of the edge device is higher than that of the main device, and the edge current is overshadowed by the main device current for all values of gate voltage, such that the presence of the edge transistor has no detrimental effect on the overall device characteristics. If, on the other hand, the P+ impurity concentration at the edges of the silicon island is too low, the threshold voltage of the edge device is lower than that of the main device, and the edge transistor current dominates the current characteristics of the overall device at low gate voltage values (Figure 4.2.2). This results in a degradation of the subthreshold characteristics and an increase of the off-state leakage current.

The mesa isolation technique is another way of isolating silicon islands from one another. This technique is attractive because of its simplicity. It simply consists into patterning the silicon into islands -or "mesas"- using a mask step and a silicon etch step. Passivation of the island edges is performed at the gate oxidation step, where the gate dielectric is grown not only on top of the silicon islands, but on their edges as well (Figure 4.2.3). This technique has been extensively used in SOS processing, where a KOH solution was used to etch the silicon and produce sloped island edges.

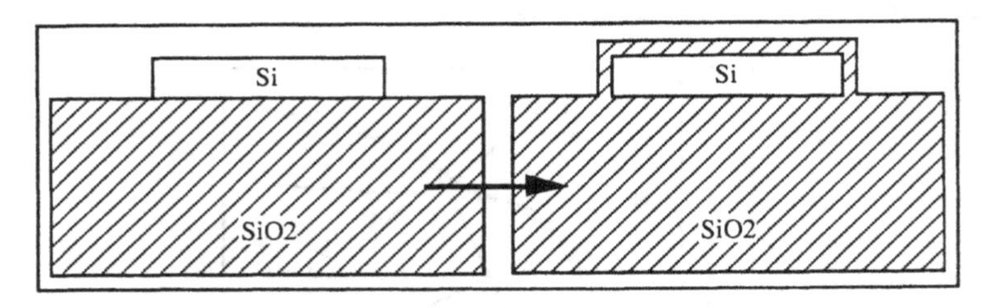

Figure 4.2.3: Mesa isolation

Several problems are associated with the mesa isolation. It is well known that the oxidation of silicon corners produces SiO_2 layers with non-uniform thickness. Indeed, the thickness of the oxide grown on the corners of an island can be 30 to 50% thinner than that grown on its top surface [4.2-4.4]. This thinning of the oxide depends on the oxidation temperature (the effect is more pronounced if oxidation is carried out below or close to the SiO_2 viscous flow temperature (965°C) [4.2]).

Oxidation is also known to sharpen silicon corners [4.4], such that corners sharper than an angle of 45 degrees are produced. This effect is enhanced if more than a single oxidation step is performed (*i.e.*: if a sacrificial oxide is grown and stripped prior to gate oxidation). The thinning of the oxide and the sharpening of the corners both contribute to a reduction of the gate oxide breakdown voltage observed when mesa isolation is used. Sidewall leakage may also be observed, as in the case where LOCOS isolation is used. In a mesa process, the gate oxide and the gate material covers both the top of the silicon island and its edges. Therefore, there exist lateral (edge) transistors in parallel to the main (top) device. Furthermore, due to charge sharing between the main and the edge devices, the threshold voltage is reduced at the corner of the island [4.5]. This can produce a kink in the subthreshold characteristics as well as leakage currents similar to those described in Figure 4.2.2. Both the oxide breakdown and the leakage currents can be improved by using P^+ sidewall doping and mesa edge rounding techniques [4.6].

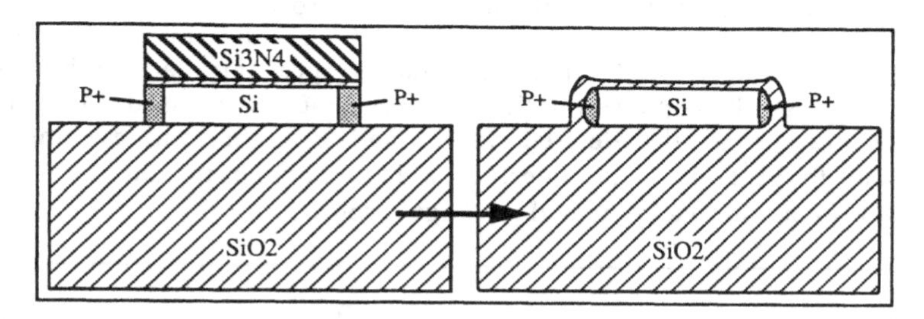

Figure 4.2.4: Oxidized mesa isolation.

A third isolation technique, called the oxidized mesa technique, results from the combination of mesa and LOCOS processes. As in the LOCOS formation, a nitride layer is patterned to define the active silicon areas. Boron is implanted around those islands which will contain n-channel devices. Some of the boron is then driven laterally in the silicon located underneath the nitride, after which the silicon is etched away to form mesas. The sidewalls of the silicon islands are then oxidized (Figure 4.2.4), and the nitride is stripped. The oxidized mesa process has several advantages. The corners of the silicon islands are rounded during the lateral oxidation step. This increases the breakdown voltage of gate oxide by over 30%, compared to the mesa isolation process [4.7]. In addition, there are no regions where the silicon is extremely thin, as it is the case when LOCOS isolation is used. This contributes to improve the subthreshold characteristics and reduce leakage currents.

4.3. Doping profiles

The optimization of the doping profile in SOI MOSFETs serves two main purposes: the adjustment of the front threshold voltage and the elimination of back-channel leakage. Different gate materials can be used. The doping profile in a device realized with a P^+ polysilicon gate will, of course be totally different from that of a device having an N^+ polysilicon as gate material [4.8-4.13]. We will only consider here the case of N^+ poly gate, which is, by far, the most common.

In "thick-film", partially depleted devices, the silicon film is thick enough ($t_{si} \geq$ ...200... nm) for the front implant controlling the front threshold voltage and the back implant controlling the back threshold voltage to be carried out independently. The dopant profile of the (enhancement-mode) n-channel transistors presents a double hump (Figure 4.3.1.A). Indeed, a superficial boron implant is used to adjust the threshold voltage to the desired value, and a higher energy implant is carried out to give the back interface a high enough threshold voltage ($\geq$ ...10... V) to avoid back-channel leakage problems.

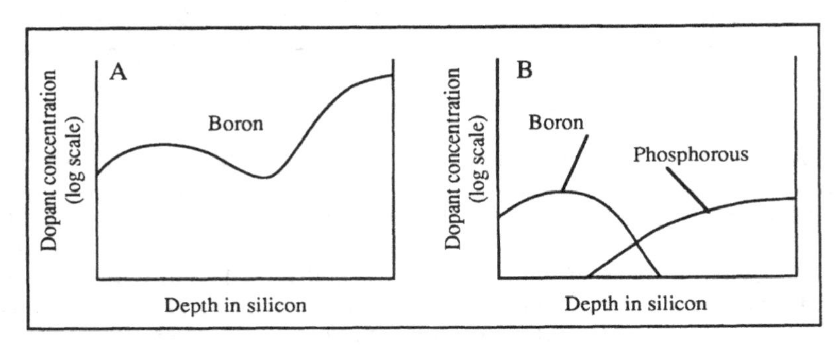

Figure 4.3.1: Doping profiles in "thick-film" SOI MOSFETs. A: n-channel device. B: p-channel device.

The doping profile of a "thick-film" p-channel transistor is shown in Figure 4.3.1.B. It is similar to the profile found in a buried-channel bulk device. A deep n-type (usually phosphorous) implant is used to avoid leakage between source and drain and to control the drain punchthrough voltage, while a shallow p-type implant is used to adjust the front threshold voltage.

In the case of thin-film ($t_{si} \leq$ ...100... nm) MOS devices, there is no room to create anything else than an almost flat doping profile (this "flat" profile is only the center portion of the gaussian-like profile produced by the implantation), and front and back threshold voltages cannot be adjusted independently. The doping profiles of thin-film n-channel and p-channel devices are presented in Figure 4.3.2. For a front gate oxide thickness of ...20.. nm, the p-type impurity concentration is ...10^{17} cm^{-3}... in the n-channel transistor and ...$4x10^{16}$ cm^{-3}... in the p-channel device. The latter operates as an accumulation-mode device, the channel region of which is fully depleted of holes (and, therefore, non-conducting) when the device is turned off. An accumulation channel forms at the top Si-SiO_2 interface when a negative gate voltage is applied. The n-channel transistor is a standard enhancement-mode device.

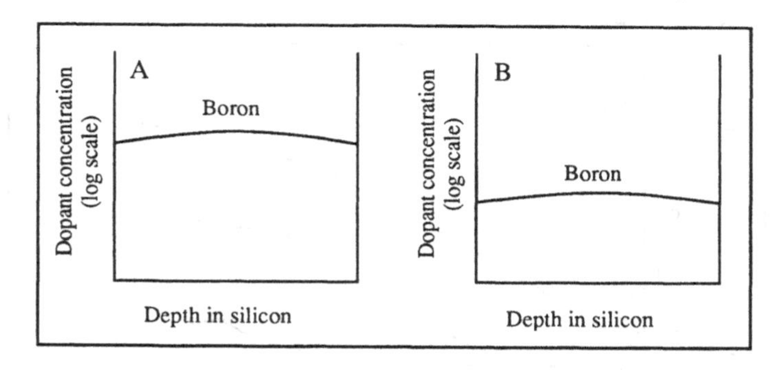

Figure 4.3.2: Doping profiles in thin-film SOI MOSFETs. A: n-channel device. B: p-channel device.

The optimization of the dopant concentration in thin-film, fully-depleted devices is a matter of balance between two effects. Firstly, the dopant concentration must be low enough to ensure full depletion, and, secondly, it must be high enough to provide the device with a suitable (*i.e.*: large enough) threshold voltage. Finding such a balance is usually not a problem in p-channel devices, but it requires some attention for n-channel transistors. Simulations show that both dopant concentration and silicon film thickness have to be optimized in order to produce useful values of threshold voltage [4.9]. Figure 4.3.3 illustrates the problem. The subthreshold slope of SOI n-channel MOSFETs can be used as a signature of the operation in the fully-depleted mode. If the slope is, say, smaller than 70 mV/decade, the device is fully depleted. If it is larger than, say, 80 mV/decade, it operates in the partially-depleted mode. The simulation of Figure 4.3.3 was carried out for a gate oxide thickness of 15 nm, a front-gate oxide charge density of 5×10^{10} cm^{-2} and a buried oxide charge density of 10^{11} cm^{-2} [4.9]. When the SOI film is relatively thick (200 nm), the device always operates in the partially-depleted mode, unless the dopant concentration is reduced to such low values that the threshold voltage is too low for useful applications (<0.1V). If the silicon film thickness is reduced to 100 nm, fully-depleted operation is maintained even if the channel dopant concentration is increased to produce threshold voltages up to 300-400 mV. Further reduction of the silicon film thickness (70 nm) allows one to reach a threshold voltage of 0.7 V while still operating in the fully-depleted mode.

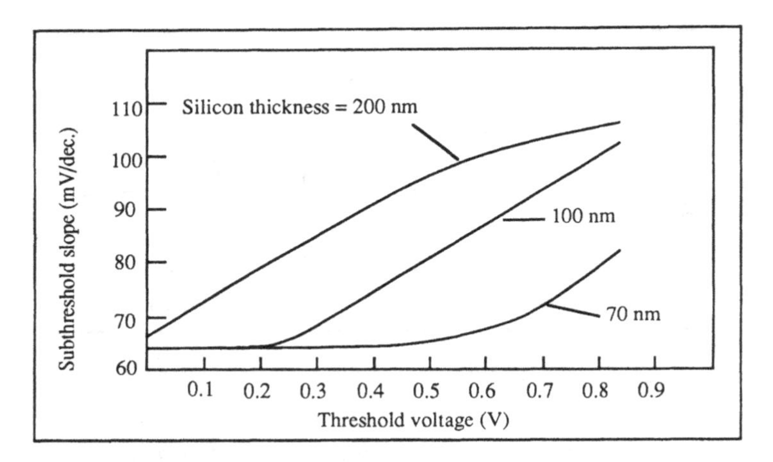

Figure 4.3.3: Subthreshold slope as a function of threshold voltage for various values of the SOI film thickness (n-channel device).

4.4. Source and drain silicidation

In thin-film SOI technology, the source and drain sheet resistance can reach high values which can jeopardize the speed performances of the circuits (the sheet resistance is roughly inversely proportional to the film thickness). Therefore, it becomes imperative to form a silicide on the sources and the drains to reduce their sheet resistance. Titanium silicide ($TiSi_2$) is the most widely used silicide in SOI technology. In bulk processes, a titanium thickness of ...60 nm... is usually deposited on the silicon in order to form the silicide. If such a "thick" titanium layer is used on thin SOI material (100 nm or less), metal-rich silicide is formed, which can lead to short-circuits between the source and drain and the gate (we consider the case of salicide (self-aligned silicide) formation). Hence, thinner titanium films have to be used to form the silicide. Figure 4.4.1 presents the source and drain sheet resistance (over N^+ and P^+ junctions) of thin-film SOI MOSFETs as a function of the deposited titanium thickness [4.10]. The silicon film thickness ranges from 65 to 125 nm. The source and drain sheet resistance can be reduced from ...300 Ω/square... to 2 Ω/square by sputtering 45 nm of titanium and forming the silicide in a two-step annealing process. The use of a thicker titanium layer led to shorts between the gate and the source and drain in the reported experiment [4.10].

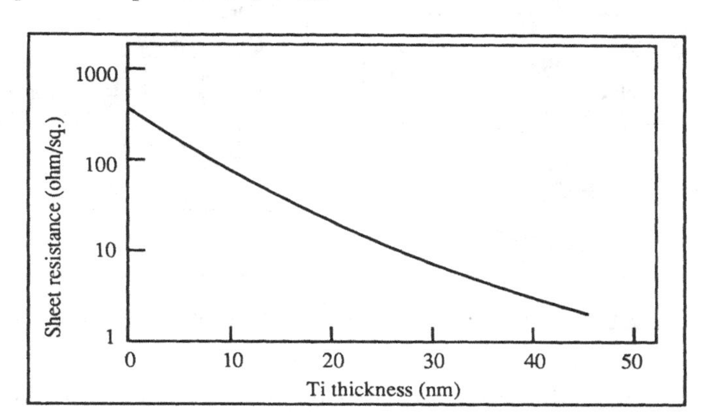

Figure 4.4.1: Source and drain sheet resistance as a function of the thickness of the deposited titanium thickness.

4.5. SOI MOSFET design

There exist different types of designs of SOI transistors. The densest and most common layout is presented in Figure 4.5.1.A. It consists of a rectangular active area, a gate, and contact holes. In the case of an n-channel device (which will be illustrated throughout this Section) the pattern of the active area is surrounded by another mask pattern (dark field if positive resist is employed) which is used for the field implant, the N-channel V_{th} adjust and back-channel stop implants, as well as for the N^+ source and drain implant steps. Similarly, the p-channel transistors are enclosed in a pattern defining the P^+ source and drain implant. The latter pattern can also be used to create a P^+ body contact in n-channel devices. When the application in which a circuit is used may lead to edge leakage problems (such as in devices submitted to ionizing radiations which can generate huge amounts of oxide charges in the oxide at the edges of the silicon islands), "edgeless" device designs can be utilized (Figure 4.5.1.B). In such a device, the silicon island (active area) presents no edge underneath the gate between the source and the drain. It is, however, worth noting that edgeless devices occupy much more silicon real estate than conventional devices, and are not used where integration density is a prime concern.

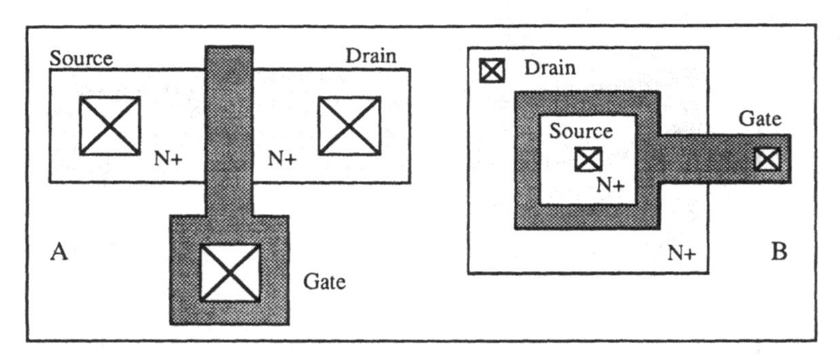

Figure 4.5.1: Layout of SOI MOSFETs. A: "normal" device. B: edgeless device.

Some applications require devices having body contacts. Indeed, contacting the silicon underneath the gate effectively suppresses the kink effect as well as parasitic lateral bipolar effects. Several schemes exist to provide the transistor body with a contact. The conventional contact is presented in Figure 4.5.2.A. It consists into a P^+ diffusion which is in contact with the P-type silicon underneath the gate. Such a device can also be used as a lateral bipolar transistor, the P^+ diffusion being the base contact, and the source and drain being used as emitter and collector, respectively. In transistors with large gate width, the presence of a single body contact at one end of the channel region may not be sufficient to suppress kink or bipolar effects. These effects can indeed take place underneath the gate, "far" from the body contact, the efficiency of which is reduced by the high resistance of the weakly doped channel region. The H-gate MOSFET design helps solving this problem, since body contacts are present at both ends of the channel (Figure 4.5.2.B). Furthermore, the H-gate device offers no direct edge leakage path between source and drain (the edges run only from N^+ to P^+ diffusions) [4.11].

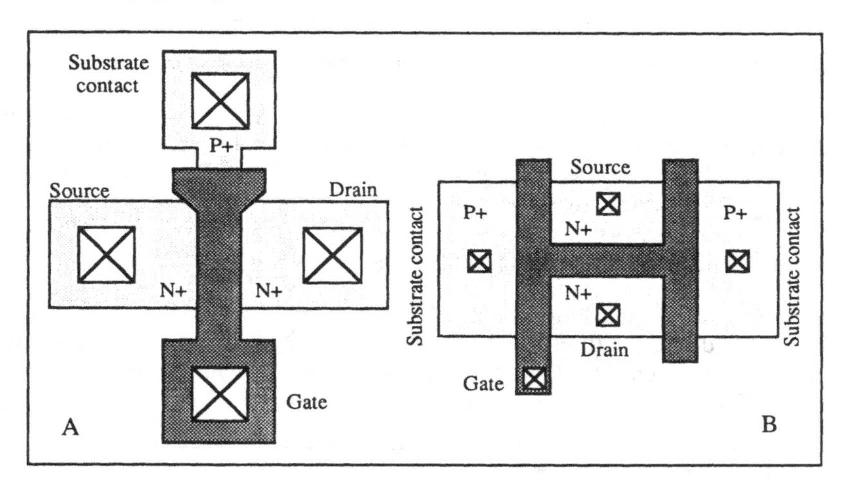

Figure 4.5.2: Transistors with body contact. A: "normal" contact. B: H-gate device.

A third type of body contact, more compact than the previous ones, is shown in Figure 4.5.3. The P+ body contacts are created on the side of the N^+ source diffusion. As in the case of the H-gate device, there is no direct edge leakage path between source and drain (the edges of the active area under the gate run only from N^+ to P^+ diffusions). If the device is very wide, additional P^+ regions can be formed in the source (such that a P^+-N^+-P^+...N^+-P^+ structure is produced). This device has the drawback of being asymmetrical (source and drain cannot be swapped), and the effective channel width, W_{eff}, is smaller than the width of the active area [4.12].

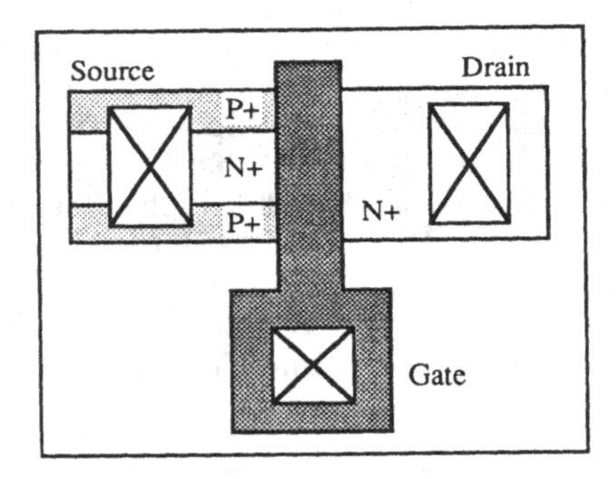

Figure 4.5.3: N-channel transistor with body contact at the source.

It is worth mentioning that body contacts are used in "thick-film", partially-depleted devices only. In thin-film devices, the full depletion gives the silicon below the gate an almost infinite resistivity which renders body contacts totally ineffective.

4.6. SOI-bulk CMOS design comparison

Generally speaking, SOI CMOS technology offers a higher integration density than bulk CMOS. This becomes evident from comparison between the layout of a bulk CMOS inverter and that of an SOI CMOS inverter (Figures 4.6.1 and 4.6.2).

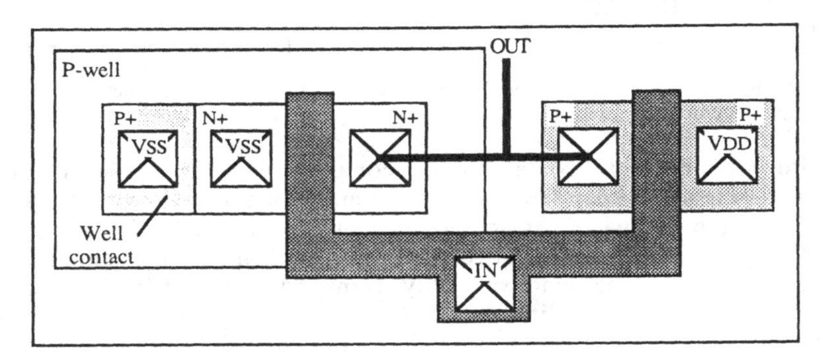

Figure 4.6.1: Layout of a bulk CMOS inverter.

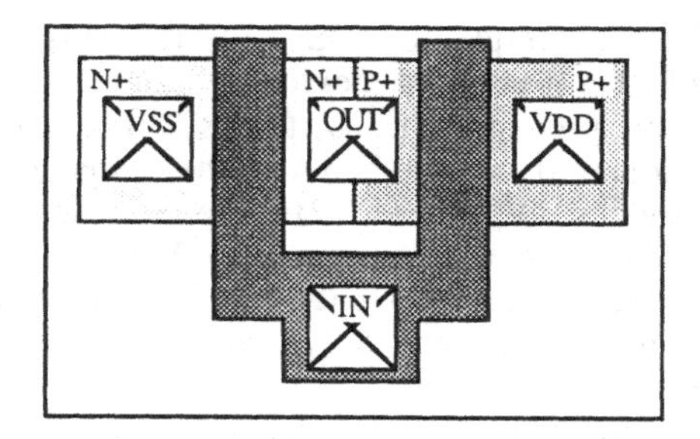

Figure 4.6.2: Layout of an SOI CMOS inverter.

This higher density results mainly from the absence of wells in SOI. A second cause of density increase is the possibility offered by SOI of having a direct contact between P^+ and N^+ junctions (such as the drains of the n-channel and the p-channel devices of Figure 4.6.2). The number of contact holes per gate is also lower in SOI than in bulk. This reduces a source of fabrication yield hazard, compared to bulk.

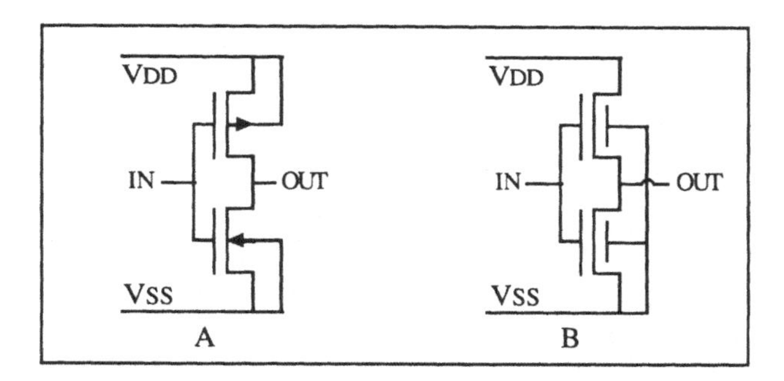

Figure 4.6.3: Back-gate (body) bias configuration in bulk (A) and SOI (B) CMOS inverters.

One of the major differences between SOI and bulk design is the difference of body effect and of body/back gate bias conditions. The body effect induced by the back gate (dV_{th1}/dV_{G2}) is negligible in partially depleted devices. The expressions for the body effect in thin-film devices (the dependence of threshold voltage on back-gate bias) can be derived from Sections 5.3.2 and 5.8 for n-channel and p-channel devices, respectively. Furthermore, the back-gate bias configuration of SOI MOSFETs is different from the substrate bias used in bulk. Let us take the example of a simple CMOS inverter (Figure 4.6.3). In bulk CMOS, the body of the n-channel device is connected to ground (V_{SS}), while the body of the p-channel transistor is connected to V_{DD} (usually +5 V). Hence, the potential of the body is the same as that of the source in both types of devices (V_{sub} = 0). In the SOI inverter, the back gate (the underlying silicon wafer) is common to both n- and p-type devices. It is usually grounded. Hence, the back- gate voltage is 0 V for the n-channel device, but it is equal to $-V_{DD}$ for the p-channel transistor, the source voltage being always used as a reference. As a consequence, SOI p-channel transistors have usually to be designed for operating with a back-gate bias, V_{G2}, equal to $-V_{DD}$ (V_{G2} = -5 V, in most cases).

CHAPTER 5 - The SOI MOSFET

Although most types of devices can be fabricated in SOI films, the preferred application field for Silicon-on-Insulator technology is undeniably CMOS. Other types of devices (bipolar devices, COMFETs, novel devices) will be reviewed in Chapter 6. SOI MOSFETs exhibit interesting properties which make them particularly attractive for applications such as rad-hard circuits, deep-submicron devices and high-temperature electronics. The properties of the SOI MOSFET operating in a harsh environment will be described in Chapter 7.

5.1. Introduction

Contrarily to bulk CMOS where both junction and field oxide isolations are used, CMOS SOI devices are dielectrically isolated from one another. This rules out latch-up between devices, as indicated in Chapter I. Similarly, there is no leakage path between devices, while surface leakage problems and field transistor action may occur in bulk technologies. Full dielectric isolation can be interesting for monolithic integration of both high-voltage devices and low-voltage CMOS on a single chip.

In bulk MOS devices, the parasitic drain (or source)-to-substrate (or well) capacitance consists of two components: the capacitance between the junction and the substrate itself, and the capacitance between the junction and the channel-stop implant under the field oxide (Figure 5.1.1.A). As devices are shrunk to smaller geometries, higher substrate dopant concentrations are used, and the areal junction capacitance increases as well. In SOI devices (with reach-through junctions), the junction capacitance has only one component: the capacitance of the MOS structure made of the junction (gate electrode of the MOS structure under consideration), the buried oxide (gate oxide of the MOS structure), and the underlying silicon substrate (substrate of the MOS structure). This parasitic capacitance can only be smaller than the capacitance of the buried oxide (Figure 5.1.1.B), which is typically lower than the capacitance junction of a bulk MOSFET. This reduction of parasitic capacitances contributed to the excellent speed performances observed in CMOS/SOI circuits. In addition, the buried oxide thickness does not have to be scaled down as smaller device geometries are reduced. This reinforces the capacitance advantage of SOI over bulk as technologies evolve towards submicron dimensions.

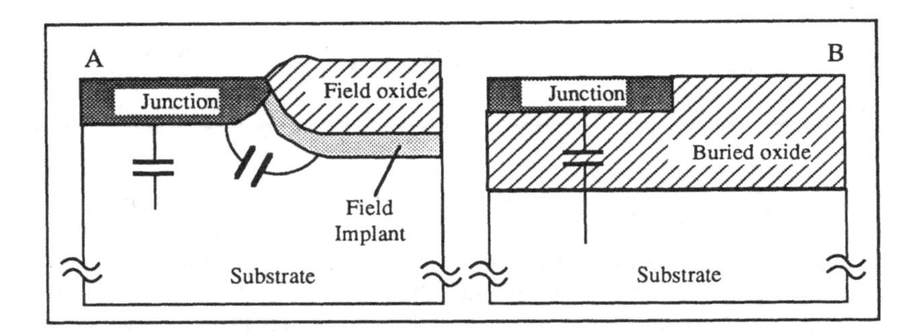

Figure 5.1.1: Parasitic junction capacitances. A: Capacitance between a junction and the substrate and between the junction and the field (channel-stop) implant in a bulk device. B: Capacitance between a junction and the substrate, across the buried oxide, in an SOI device.

The presence of a buried oxide underneath the devices reduces not only the junction capacitances, but some other capacitances as well (all the capacitances between the silicon substrates and another terminal). Table 5.1.1. presents typical capacitances of a bulk and an SOI 1-micrometer processes [5.45]. The capacitances are given in pF/μm^2. The reduction of capacitance is, of course, most noticeable between the junctions and the substrate, but one can observe that even the metal 1-to-substrate capacitance can be reduced by 40% by using SOI substrates rather than bulk silicon wafers.

Capacitor type	SOI (SIMOX)	Bulk	Gain (SOI vs. bulk)
Gate	1.3	1.3	1
Junction-to-substrate	0.05	0.2 ... 0.35	4 ... 7
Polysilicon-to-substrate	0.04	0.1	2.5
Metal 1-to-substrate	0.027	0.05	1.85
Metal 2-to-substrate	0.018	0.021	1.16

Table 5.1.1: Parasitic capacitances (pF/μm^2) found in typical bulk and SOI 1-μm CMOS processes [5.45].

5.2. Distinction between thick- and thin-film devices

All SOI MOSFETs are not alike. Their physics is highly dependent on the thickness of the silicon film in which they are made. Three types of devices can be distinguished, depending on both the silicon film thickness and the channel doping concentration: the thick-film and the thin-film devices, as well as the "medium thickness" device, which can exhibit either a thin- or a thick-film behavior, depending on the back-gate bias. Figure 5.2.1 presents the band curvature diagrams of a bulk, a thick-film SOI, and a thin-film SOI n-channel device at threshold.

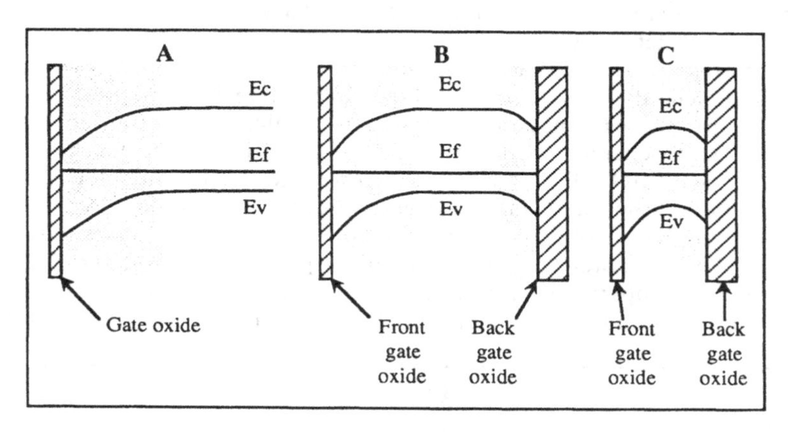

Figure 5.2.1: Band diagram in a bulk (A), a thick-film SOI (B), and a thin-film SOI device. All devices are represented at threshold (front gate voltage = threshold voltage). The shaded areas represent the depleted zones. SOI devices are represented for a condition of weak inversion (below threshold) at the back interface.

In the **bulk** device (Figure 5.2.1.A), the depletion zone extends from the Si-SiO_2 interface down to the maximum depletion width, x_{dmax}, which is classically given by $\sqrt{\frac{4\varepsilon_{si}.\Phi_F}{q\,N_a}}$, Φ_F being the Fermi potential, which is equal to $\frac{kT}{q}\ln\left(\frac{N_a}{n_i}\right)$.

In a **thick-film SOI** device (Figure 5.2.1.B), the silicon film thickness is larger than twice the value of x_{dmax} . In such a case, there is no interaction between the depletion zones arising from the front and the back interfaces, and there exists a piece of neutral silicon beneath the front depletion zone. If this neutral piece of silicon, called "body", is connected to ground by a "body contact", the characteristics of the device will exactly be those of a bulk device. If the body is left electrically floating, the device will basically behave as a bulk device, with the notable exception of two parasitic effects, the first of which is called "kink effect" (see Section 5.6), the second one being the presence of a parasitic, open-base NPN bipolar transistor between source and drain (see Section 5.8).

In a **thin-film SOI** device (Figure 5.2.1.C), the silicon film thickness is smaller than x_{dmax} . In that case, the silicon film is fully depleted at threshold, irrespective of the bias which is applied to the back gate (with the exception of the possible presence of thin accumulation or inversion layers at the back interface, if a large negative or positive bias is applied to the back gate, respectively). Thin-film, fully depleted SOI devices are virtually free of kink effect, if their back interface is not in accumulation. Among all types of SOI devices, fully depleted devices with depleted back interface exhibit the most attractive properties, such as low electric fields, high transconductance, excellent short-channel behavior, and a quasi-ideal subthreshold slope. Thin-film SOI MOSFETs are often referred to as fully-depleted devices. Because both front and back interfaces can be in either accumulation, depletion or inversion, one can number 9 modes of operation in the thin-film SOI transistor as a function of V_{G1} and V_{G2} (Figure 5.2.2) [5.1]. Most of these operating modes are not of practical use, however. The useful operation modes are indicated by the shaded area of Figure 5.2.2.

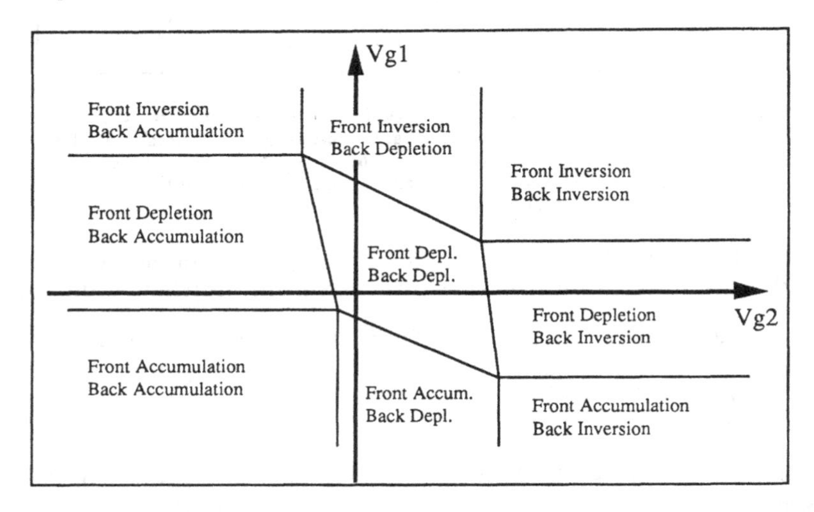

Figure 5.2.2: Different operating modes of a thin-film SOI n-channel MOS transistor as a function of front-gate bias (V_{G1}) and back-gate bias (V_{G2}) - (linear regime, low drain voltage). The shaded area represents the normal mode of operation.

To be more general, it should be mentioned that the presence of accumulation, depletion or inversion layers is also a function of the drain voltage, and that the back interface can, for instance, be accumulated near the source and depleted near the drain. Such a mode of operation will be analyzed later on. It also worthwhile noting that most of the "attractive properties" of fully depleted devices with depleted back interface (high transconductance,...) can be disabled by the presence of an accumulation layer at the back interface.

The previous remarks are valid for enhancement-mode MOSFETs, which constitute the most popular class of SOI devices (at least in the n-channel case). It is worth noting, however, that another type of devices can be realized. These are accumulation-mode (or "deep-depletion") devices. The operation mode of a MOSFET (enhancement or accumulation) depends on fabrication parameters, the most important being the type of gate material used [5.2]. We will, however consider that thin-film SOI devices are enhancement-mode devices, unless otherwise specified.

	N+ poly gate	P+ poly gate
n-channel device	Inversion	Accumulation
p-channel device	Accumulation	Inversion

Table 5.2.1: Different operation modes of a thin-film SOI MOSFET as a function of gate material.

Medium-thickness SOI devices are an intermediate case between thick- and thin-film devices, and are obtained in those cases where $x_{dmax} < t_{si} < 2x_{dmax}$, t_{si} being the silicon film thickness. If the back-gate bias is such that the front and back depletion zones do not touch each other, or if the back interface is neutral or accumulated, the transistor will behave as a thick-film device. If, on the other hand, the presence of a back-gate bias induces coalescence of the front and back depletion zones, the device will be fully depleted, and it will behave as a thin-film device.

The merits of the different types of SOI MOSFETs are reported in Table 5.2.2, where some electrical properties of the devices are compared. Bulk silicon devices are taken as a reference. One can see that thin-film, fully depleted devices without accumulation at the back interface offer the most attractive properties for ULSI applications. The popularity of thicker, partially depleted devices is due to a property which is not listed in this table, and which is the superior performances in a harsh environment.

	Bulk	Thick-film SOI	Thin-film SOI Back accum.	Thin-film SOI Fully depleted
Mobility	0	0	0/-	+
Transconductance	0	0	0/-	+
Short-channel effect	0	0	+	0/+
S&D capacitance	0	+	+	+
Hot carriers	0	0/+	0/-	+
Subthreshold slope	0	0	0/-	+
Vth sensitivity on tsi	0	0	-	-
Kink	0	-	-	0
Parasitic bipolar	0	-	-	0/-
Total-dose hardness	0	0/+	0/-	-
SEU hardness	0	+	+	+
Soft-error hardness	0	+	+	+

Table 5.2.2: Comparison of some of the electrical properties of thick-film, thin-film with accumulated back interface, and thin-film, fully depleted SOI devices. The bulk device is given as a reference. 0,+ and - mean "similar to bulk", "better than bulk" and "worse than bulk", respectively.

The following section of this chapter will describe the properties and the characteristics of thick- and thin-film MOSFETs. Medium-thickness devices behave in a rather complex way and can be treated as either thick- or thin-film devices, as a function of back-gate, front-gate or drain bias conditions. Indeed, in such a device the back-interface charge condition, for instance, can vary between accumulation, neutrality and depletion along the device length from source to drain.

5.3. I-V Characteristics

5.3.1. Threshold voltage

The threshold voltage of an enhancement-mode **bulk** n-channel MOSFET is classically given by [5.3] :

$$V_{th} = V_{FB} + 2\Phi_F + \frac{q\, N_a\, x_{dmax}}{C_{ox}} \qquad (5.3.1)$$

where V_{FB} is the flatband voltage, equal to $\Phi_{MS} - \frac{Q_{ox}}{C_{ox}}$ (we will neglect the presence of fast surface states, N_{it}, in the present analysis), Φ_F is the

Fermi potential, equal to $\frac{kT}{q} \ln\left(\frac{N_a}{n_i}\right)$, and x_{dmax} is the maximum depletion width, equal to $\sqrt{\frac{4\varepsilon_{si}\Phi_F}{q N_a}}$.

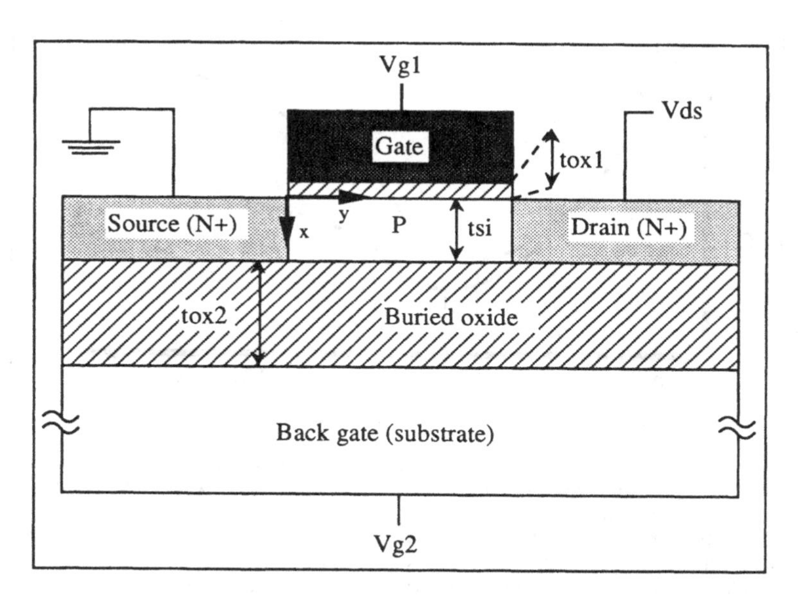

Figure 5.3.1: Cross-section of at thin-film, n-channel SOI MOSFET illustrating some of the notations used in this section.

In a **thick-film** SOI device ($t_{si} > 2\ x_{dmax}$), there can be no interaction between the front and back depletion zones. In that case, the threshold voltage is the same as in a bulk transistor and is given by equation (5.3.1).

The threshold voltage of a **thin-film**, fully depleted, enhancement-mode n-channel SOI device [5.4] (Figure 5.3.1) can be obtained by solving the Poisson equation using the depletion approximation: $\frac{d^2\Phi}{dx^2} = \frac{q N_a}{\varepsilon_{si}}$ which can be integrated twice and yields the potential as a function as the depth in the silicon film, *x*:

$$\Phi(x) = \frac{q N_a}{2\ \varepsilon_{si}} x^2 + \left(\frac{\Phi_{s2}-\Phi_{s1}}{t_{si}} - \frac{q N_a t_{si}}{2\ \varepsilon_{si}}\right) x + \Phi_{s1} \qquad (5.3.2)$$

where Φ_{s1} and Φ_{s2} are the potentials at the front and back silicon/oxide interfaces, respectively (Figure 5.3.2). The dopant concentration, N_a, is assumed to be constant.

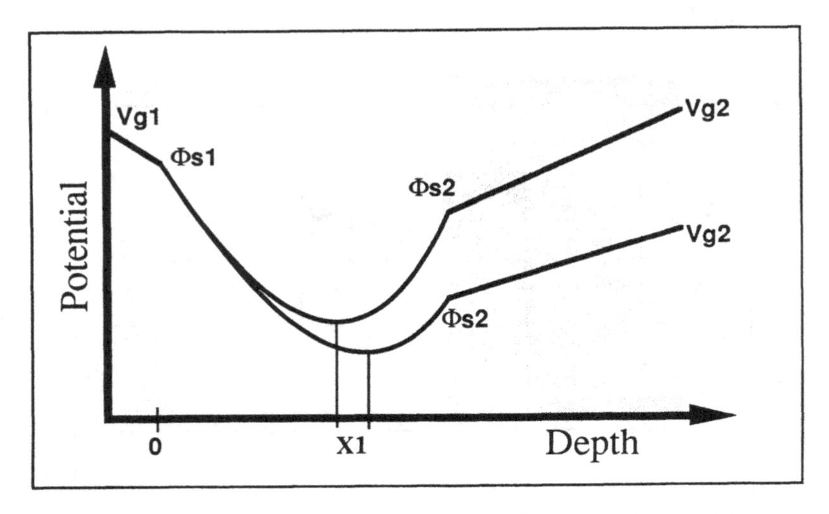

Figure 5.3.2: Potential in the silicon film, front and back oxides, at $V_{g1}=V_{th1}$, and for two back-gate bias conditions. x_1 is the point of minimum potential. The depletion zone between x=0 and $x=x_1$ is controlled by the front gate, and the depletion zone between $x=x_1$ and the back $Si\text{-}SiO_2$ interface is controlled by the back gate. The shaded areas represent the gate (left) and buried (right) oxides.

The electric field in the silicon film is given by:

$$E(x) = \frac{-q\,N_a}{\varepsilon_{si}}\,x - \left(\frac{\Phi_{s2}-\Phi_{s1}}{t_{si}} - \frac{q\,N_a\,t_{si}}{2\,\varepsilon_{si}}\right) \tag{5.3.3}$$

The front surface electric field, E_{s1} (at x=0), can be calculated from (5.3.3) and is given by:

$$E_{s1} = \left(\frac{\Phi_{s1}-\Phi_{s2}}{t_{si}} + \frac{q\,N_a\,t_{si}}{2\,\varepsilon_{si}}\right) \tag{5.3.4}$$

Applying Gauss' theorem at the front interface, one obtains the potential drop across the gate oxide, Φ_{ox1}:

$$\Phi_{ox1} = \frac{\varepsilon_{si}\,E_{s1} - Q_{ox1} - Q_{inv1}}{C_{ox1}} \tag{5.3.5}$$

where Q_{ox1} is the fixed charge density at the front Si-SiO_2 interface, Q_{inv1} is the front channel inversion charge ($Q_{inv1} < 0$), and C_{ox1} is the front gate oxide capacitance. Similarly, applying Gauss' theorem at the back interface and using (5.3.4) yields the potential drop across the buried oxide, Φ_{ox2}:

$$\Phi_{ox2} = -\frac{\varepsilon_{si} E_{s1} - q N_a t_{si} + Q_{ox2} + Q_{s2}}{C_{ox2}} \tag{5.3.6}$$

where Q_{s2} is the charge in a possible back inversion ($Q_{s2} < 0$) or accumulation ($Q_{s2} > 0$) layer.

The front and back gate voltages, V_{G1} and V_{G2} , are given by:

$$V_{G1} = \Phi_{s1} + \Phi_{ox1} + \Phi_{MS1} \quad \text{and} \quad V_{G2} = \Phi_{s2} + \Phi_{ox2} + \Phi_{MS2} \tag{5.3.7}$$

where Φ_{MS1} , Φ_{MS2} are the front and back work function differences, respectively.

By combining (5.3.4), (5.3.5) and (5.3.7), we obtain the relationship between the front gate voltage and the surface potentials:

$$V_{G1}=\Phi_{MS1} - \frac{Q_{ox1}}{C_{ox1}} + \left(1 + \frac{C_{si}}{C_{ox1}}\right)\Phi_{s1} - \frac{C_{si}}{C_{ox1}}\Phi_{s2} - \frac{\frac{1}{2} Q_{depl} + Q_{inv1}}{C_{ox1}} \tag{5.3.8}$$

where $C_{si} = \varepsilon_{si}/t_{si}$ and Q_{depl} is the depletion charge in the silicon film, which is equal to $-q N_a t_{si}$.

Similarly, one finds the relationship between the back gate voltage and the surface potentials:

$$V_{G2}=\Phi_{MS2} - \frac{Q_{ox2}}{C_{ox2}} - \frac{C_{si}}{C_{ox2}}\Phi_{s1} + \left(1 + \frac{C_{si}}{C_{ox2}}\right)\Phi_{s2} - \frac{\frac{1}{2} Q_{depl} + Q_{s2}}{C_{ox2}} \tag{5.3.9}$$

Equations (5.3.8) and (5.3.9) are the key relations which describe the charge coupling between the front and back gates in a fully depleted SOI MOSFET [5.4]. Combining them yields the dependence of the (front) threshold voltage on back-gate bias and device parameters.

We will now detail the expression of the threshold voltage of the thin-film SOI MOSFET as a function of the different possible steady-state charge conditions at the back interface [5.4].

If the back surface is **accumulated**, Φ_{s2} is pinned to approximately 0V. The threshold voltage, $V_{th1,acc2}$ is, therefore, obtained from equation (5.3.8) where $V_{th1,acc2} = V_{G1}$ is calculated at $\Phi_{s2} = 0$, $Q_{inv1}=0$, and $\Phi_{s1} = 2\Phi_F$. The result is:

$$V_{th1,acc2} = \Phi_{MS1} - \frac{Q_{ox1}}{C_{ox1}} + \left(1 + \frac{C_{si}}{C_{ox1}}\right) 2\Phi_F - \frac{Q_{depl}}{2C_{ox1}} \qquad (5.3.10)$$

If the back surface is **inverted**, Φ_{s2} is pinned to approximately $2\Phi_F$. The front threshold voltage, $V_{th1,inv2}$ is, therefore, obtained from equation (5.3.8) where $V_{th1,inv2} = V_{G1}$ is calculated at $\Phi_{s2} = 2\Phi_F$, $Q_{inv1}=0$, and $\Phi_{s1} = 2\Phi_F$. The result is:

$$V_{th1,inv2} = \Phi_{MS1} - \frac{Q_{ox1}}{C_{ox1}} + 2\Phi_F - \frac{Q_{depl}}{2C_{ox1}} \qquad (5.3.11)$$

It is worth noting that in this case, the device is still ON even if $V_{G1} < V_{th1,inv2}$ since the back interface is inverted, and that the device is, therefore, useless for any practical circuit application.

If the back surface is depleted, Φ_{s2} depends on the back-gate voltage, V_{G2}, and its value can range between 0 and $2\Phi_F$. The value of back-gate voltage for which the back interface reaches accumulation (the front interface being at threshold), $V_{G2,acc}$ is given by equation (5.3.9) where $\Phi_{s1} = 2\Phi_F$, $\Phi_{s2} = 0$, and $Q_{s2}=0$. Similarly, the value of back-gate voltage for which the back interface reaches inversion, $V_{G2,inv}$ is given by the same equation where $\Phi_{s1} = 2\Phi_F$, $\Phi_{s2} = 2\Phi_F$, and $Q_{s2}=0$. When $V_{G2,acc} < V_{G2} < V_{G2,inv}$ the front threshold voltage is obtained by combining equations (5.3.8) and (5.3.9) with $\Phi_{s1} = 2\Phi_F$ and $Q_{inv1} = Q_{s2} =0$. The result is:

$$V_{th1,depl2} = V_{th1,acc2} - \frac{C_{si}\, C_{ox2}}{C_{ox1}\,(C_{si} + C_{ox2})} (V_{G2} - V_{G2,acc}) \qquad (5.3.12)$$

The above relationships [5.4] are valid if the thickness of the inversion or accumulation layers are small with respect to the silicon film thickness. This may no longer be the case in ultra-thin-film devices, in which case the width of the accumulation/inversion zones must be subtracted from the silicon film thickness to obtain an effective silicon thickness (= the effective width of the depleted layer) to be used in place of t_{si} in the above relationships, in a first approximation. The exact analysis of very thin films is rather complex (it involves quantum mechanical considerations) and will not be described here.

The influence of the back-gate bias on the $I_D(V_{GS})$ characteristics of a thin-film SOI n-channel MOSFET with low V_{DS} are shown in Figure 5.3.3. In the left portion of the graph (region A), the back interface is inverted. As a consequence, current flows in the device even if the front-gate voltage is negative. Φ_{s2} is pinned to $2\Phi_F$, and the front threshold voltage is fixed at a constant value. The apparent shift of the curves to the left with increased back bias is actually a shift of the curves upwards due to the increase of the back channel current. In the right portion of the graph (region C), the back interface is accumulated, the front threshold voltage is constant, and no shift of the curves to the right can be obtained by further negative increase of the back bias. In region B, the back interface is depleted, and the front threshold voltage depends linearly on the back bias. The anomalous slope found in region *a* is explained by the following effect. As the back and front interfaces are depleted, and as the back interface is close to inversion, any increase of V_{G1} will push the point of minimum potential deeper into the silicon film, and therefore reduce the back threshold voltage. This can lead to creation of an inversion channel at the back interface. Thus, one observes a situation where a variation of the bias in the front gate creates and modulates an inversion channel at the back interface [5.4, 5.5], which nicely illustrates the importance of the interaction between front and back gate in thin SOI devices.

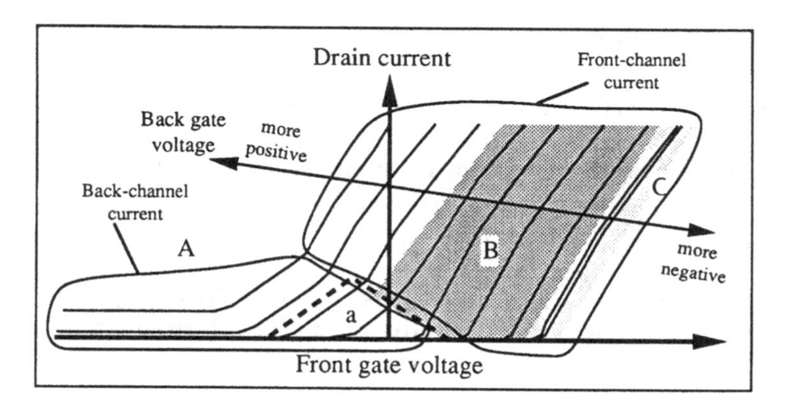

Figure 5.3.3: Linear ID(VG1) characteristics of a thin-film SOI n-channel MOSFET for different VG2 values. Different back interface conditions are outlined by the shaded areas: inversion (A), depletion (B), accumulation (C), and depletion/inversion regime depending on the front gate voltage (a).

5.3.2. Body effect

In a bulk device, the body effect is defined as the dependence of the threshold voltage on the substrate bias. In an SOI transistor, it is similarly defined as the dependence of the threshold voltage on the back-gate bias.

In a **bulk** n-channel transistor, the threshold voltage can be written as [5.6]:

$$V_{th} = \Phi_{MS} + 2\Phi_F - \frac{Q_{ox}}{C_{ox}} + \frac{Q_b}{C_{ox}} \quad \text{with} \quad Q_b = \sqrt{2\varepsilon_{si}\, q\, N_a\, (2\Phi_F - V_B)}$$

which can be rewritten:

$$V_{th} = \Phi_{MS} + 2\Phi_F - \frac{Q_{ox}}{C_{ox}} + \frac{\sqrt{2\varepsilon_{si}\, q\, N_a\, (2\Phi_F - V_B)}}{C_{ox}}$$

and, by defining $\gamma = \frac{\sqrt{2\varepsilon_{si}\, q\, N_a}}{C_{ox}}$

one obtains: $V_{th} = \Phi_{MS} + 2\Phi_F - \frac{Q_{ox}}{C_{ox}} + \gamma\sqrt{2\Phi_F} + \gamma\left(\sqrt{2\Phi_F - V_B} - \sqrt{2\Phi_F}\right)$

The last term depicts the dependence of threshold voltage on substrate bias (body effect). When a negative bias is applied to the substrate (with respect to the source), the threshold voltage increases as a square-root function of the substrate bias. If the threshold voltage with zero substrate bias is referred to as V_{th0}, one can write:

$$V_{th}(V_B) = V_{th0} + \gamma\left(\sqrt{2\Phi_F - V_B} - \sqrt{2\Phi_F}\right)$$

γ is called the body-effect parameter (unit: $V^{1/2}$).

The "body effect" (or, more accurately, the back-gate effect) can be neglected (γ=0) in **thick-film** SOI devices, because there is no coupling between front and back gate.

In a **thin-film**, fully depleted SOI device, the "body effect" (or, more accurately, the back-gate effect) can be obtained from equation (5.3.12):

$$\frac{dV_{th1}}{dV_{G2}} = -\frac{C_{si}\, C_{ox2}}{C_{ox1}\,(C_{si} + C_{ox2})} = \frac{-\varepsilon_{si}\, C_{ox2}}{C_{ox1}\,(t_{si}\, C_{ox2} + \varepsilon_{si})} = \gamma \quad (5.3.13)$$

The symbol γ is chosen by analogy with the case of a bulk device. It should be noted that γ is dimensionless in the case of thin-film SOI transistors, and that the threshold dependence on back-gate bias is

linear. In most cases, the following approximation can be made: $\gamma \cong -\frac{t_{ox1}}{t_{ox2}}$
Expression (5.3.13) is valid when the film is fully depleted only. In a first order approximation, one can consider that Φ_{S2} is pinned at $2\Phi_F$ when the back interface is inverted, and that further increase of the back-gate bias will no longer modify the front threshold voltage ($\gamma \cong 0$). Similarly, when a large negative back-gate bias is applied, the back interface is accumulated, Φ_{S2} is pinned at 0 V, and further negative increase of the back-gate bias will not modify the front threshold voltage ($\gamma \cong 0$). Variation of the front gate voltage with back-gate bias, based on these assumptions, is represented on Figure 5.3.4 [5.4]. In a real case, however, the back surface potential can exceed $2\Phi_F$ (back inversion) or become smaller than 0V (back accumulation), these excursions being limited to a few $\frac{kT}{q}$. As a result, the front threshold voltage still shows some increase (decrease) when the back-gate voltage is increased beyond the threshold of back accumulation (inversion) [5.1].

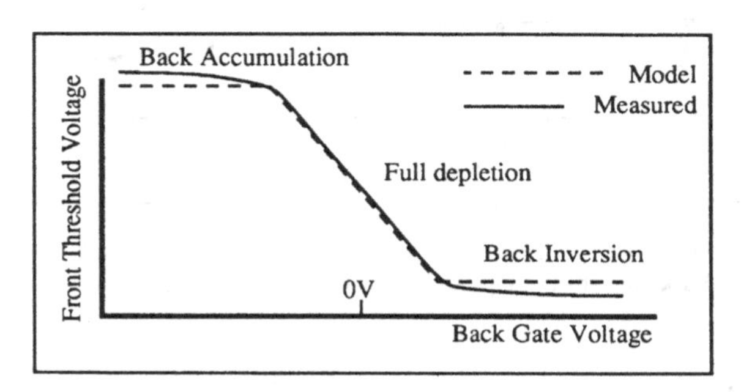

Figure 5.3.4: Variation of threshold voltage with back-gate bias.

It is worth noting that relationship (5.3.13) is independent of the doping concentration, N_a. If C_{ox1} and C_{ox2} are known, (5.3.13) can be used to determine the thickness of the silicon film [5.7] The dependence of front threshold voltage on back-gate bias decreases with increasing t_{ox2}. When t_{ox2} is very thick ($C_{ox2} \cong 0$), the front threshold voltage is virtually independent of the back-gate bias. In real SOI devices, the back-gate material is not a metal, as considered in this first-order model, but a silicon substrate, the surface of which can become inverted, depleted or accumulated as a function of the back-gate bias conditions. This variation of the substrate surface potential has some influence on the device threshold voltage, but this influence is small, and can be neglected as long as the thickness of the buried oxide layer is large compared to that of the front gate oxide.

Although direct comparison between the "body-effect" parameters γ of bulk and thin-film SOI devices cannot be made because of their different units ($V^{1/2}$ for γ_{bulk}, while γ_{SOI} has no dimension), it is clear that the back-gate effect of SOI devices is much smaller than the body effect of bulk MOSFETs. Figure 5.3.5 presents the experimental variation of threshold voltage in a bulk and a thin-film, fully depleted SOI n-channel transistor. When a -8V back (substrate) bias is applied to the bulk device, a 1.2V increase of threshold voltage is observed. Under the same bias condition, the threshold voltage of the SOI device increases only by 0.2V.

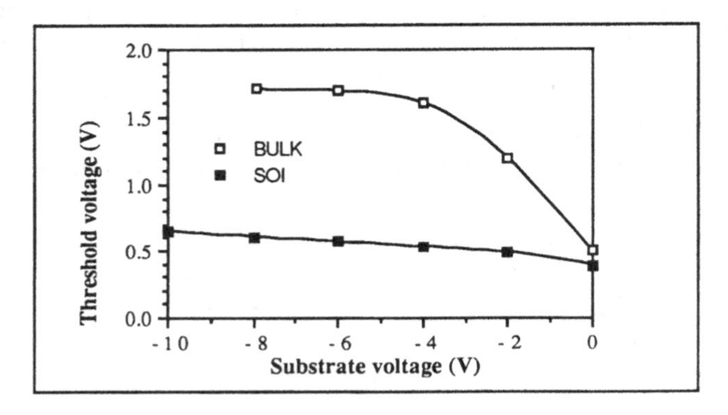

Figure 5.3.5: Dependence of threshold voltage on back bias (body effect) in bulk and fully depleted SOI MOSFETs [5.8].

The reduced body effect is an important feature of SOI devices. Indeed, the body effect reduces the current drive capability of transistors whose source is not directly connected to ground, such as transfer gates, nMOS load devices, and differential input pairs. A higher gain can, therefore, be expected from SOI gates than from their bulk counterparts.

5.3.3. Short-channel effects

There are numerous effects caused by the reduction of channel length in MOSFETs [5.16]. In this Section we will more specifically deal with the so-called "short-channel effect" which consists into a roll-off of the threshold voltage in short-channel devices. It is due to the loss of control by the gate of a part of the depletion zone below it. In other words, the depletion charge controlled by the gate is no longer equal to $Q_{depl} = \frac{q\ N_a\ x_{dmax}}{C_{ox}}$ (bulk MOSFET case), but to a fraction of it, which we will call Q_{d1}. This reduction of the depletion charge, due to the encroachment from the source and drain, becomes significant in short-channel devices, and brings about a lowering of threshold voltage obtained by substituting Q_{d1} to Q_{depl} in equation (5.3.1). In a bulk MOSFET, Q_{d1} can be geometrically represented by the area of a trapezoid (Figure 5.3.6). In a long-channel device, the lengths of the upper and the lower base of the trapezoid are almost equal to L, the channel length. In a short-channel device, the upper base length is still equal to L, but the lower base is significantly shorter (it can even disappear, as in Figure 5.3.6). The value of Q_{d1} can be approximated by [5.9]:

$$Q_{d1} = Q_{depl}\left(1 - \frac{r_j}{L}\left(\sqrt{1 + \frac{2x_{dmax}}{r_j}} - 1\right)\right)$$

where r_j is the source and drain junction depth.

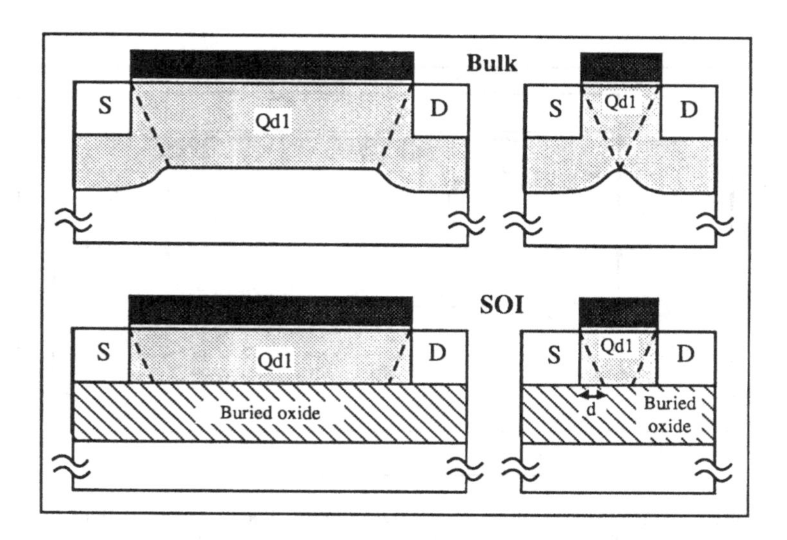

Figure 5.3.6: Distribution of depletion charges in long-channel (left) and short-channel (right) bulk and thin-film SOI MOSFETs. Q_{d1} is the depletion charge controlled by the gate.

In a thin-film SOI device, the depletion charge controlled by the gate is given by [5.10]: $Q_{d1} = Q_{depl}\left(1 - \frac{d}{L}\right)$ where d is a distance defined by Figure 5.3.6 and $Q_{depl} = q\ N_a\ t_{si}$. Calculating the value of d is rather complex and necessitates iterative calculation of the back-surface potential. A description of the calculation method is given in [5.10].

A much cruder, empirical but reasonably useable expression of Q_{d1} can be obtained for fully-depleted MOSFETs using the following approximation, adapted from [5.11] (Figure 5.3.7).

Q_{d1} is given by the trapezoid area: $Q_{d1}= Q_{depl}\left(1 - \alpha \ \frac{d_S+d_D}{2L}\right)$ where $Q_{depl}= q\ N_a\ x_1$, $\alpha = x_1/x_{dmax}$, d_S and d_D are the bases of the triangular charges controlled by source and drain in a virtual bulk device having the same doping concentration, N_a, as the SOI MOSFET under consideration. x_1 is the depth of minimum potential in the silicon film (see Figure 5.3.2), and is equal to $\frac{t_{si}}{2} + \frac{\varepsilon_{si}}{qN_a t_{si}}(2\Phi_F - \Phi_{s2})$. The values of d_S and d_D can be approximated by the depletion depths generated by the source and drain the junctions:

$$d_S = \sqrt{\frac{2\varepsilon_{si}}{qN_a}(E_g/2 + \Phi_F)} \quad \text{and} \quad d_D = \sqrt{\frac{2\varepsilon_{si}}{qN_a}(E_g/2 + \Phi_F + V_{DS})}$$

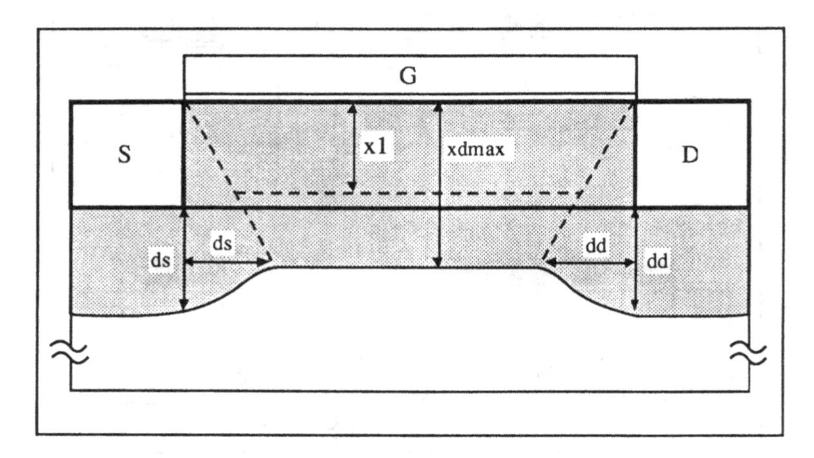

Figure 5.3.7: Cross section of an SOI MOSFET and the equivalent bulk device, used for approximating the short-channel effect.

This simplified model has been used to plot Figure 5.3.8 by replacing qN_ax_{dmax} by Q_{d1} in equation (5.3.1). One can see that threshold voltage roll-off starts to occur at significantly smaller gate lengths in thin SOI transistors than in bulk devices (this can also be seen in Figure 5.3.6, since Q_{d1} retains a reasonable trapezoid shape in the short-channel SOI device, while it shows a triangular shape ($Q_{d1} \cong \frac{Q_{depl}}{2}$) in the bulk case.

The short-channel effect is smaller in thin-film devices with accumulation at the back side than in thin-film, fully depleted devices, but both of them show less short-channel effect than bulk devices [5.13, 5.16]. It seems that an optimum control of the space charge in the silicon film by the gate (which would further minimize the short-channel effect) could be obtained by using double-gate devices (one gate below the active silicon film and one above it) [5.14]. Making such devices is a real technological challenge, but some practical solutions have already been proposed [5.15].

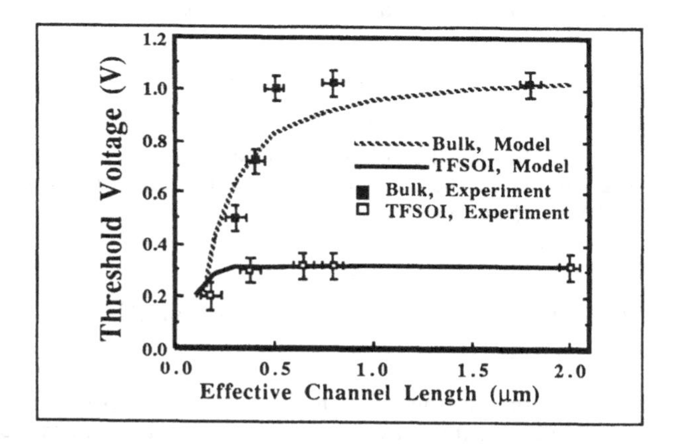

Figure 5.3.8: Threshold.voltage as a function of gate length in bulk and thin-film (100 nm) SOI n-channel MOSFETs [5.12].

Another short-channel effect, called drain-induced conductivity enhancement (DICE) is also due to charge sharing between the gate and the junctions [5.10]. DICE is caused by the reduction of the depletion charge controlled by the gate due to a size increase of the drain junction-related depletion zone, which itself increases with V_{DS}. In order to model the DICE effect, an analysis involving the solution of the two-dimensional Laplace equation (in a fully depleted device) must be carried out. It can be shown that DICE is lower in thin-film SOI devices than in bulk devices [5.16].

5.3.4. Output characteristics

The expression of the current characteristics $I_D(V_{G1}, V_{G2}, V_{DS})$ of a **thick-film** SOI MOS transistor is identical to that of a bulk MOSFET, with the expression of the kink effect and the presence of parasitical bipolar effects, which will be discussed later in this Chapter.

Derivation of the current characteristics of a **thin-film**, fully depleted SOI device can be made using assumptions of the classical gradual-channel approximation [5.3]: constant mobility as a function of y, uniform doping of the silicon film in the channel region, and negligible diffusion current. Once again, we will consider the case of an n-channel device [5.17]. A more complete analysis, including short-channel effects can be found in Ref. [5.10]. Using Ohm's law in an elemental section of the inversion channel, one can write:

$$I_D = -W\,\mu_n\,Q_{inv1}(y)\,\frac{d\Phi_{s1}(y)}{dy} \qquad (5.3.14)$$

where W is the width of the channel, and μ_n is the mobility of the electrons in the inversion layer. Integration of the previous expression from source (y=0) to drain (y=L) yields:

$$I_D = -\frac{W}{L}\,\mu_n \int_{2\Phi_F}^{2\Phi_F+V_{DS}} Q_{inv1}(y)\,d\Phi_{s1}(y) \qquad (5.3.15)$$

Assuming full depletion in the silicon film and assuming that the width of the inversion and accumulation layers at the interfaces, one obtains the inversion charge density in the front channel, $Q_{inv1}(y)$, from equation (5.3.8):

$$-Q_{inv1}(y)=C_{ox1}\left(V_{G1}-\Phi_{MS1}+\frac{Q_{ox1}}{C_{ox1}}-\left(1+\frac{C_{si}}{C_{ox1}}\right)\Phi_{s1}(y)+\frac{C_{si}}{C_{ox1}}\Phi_{s2}(y)+\frac{Q_{depl}}{2C_{ox1}}\right) \qquad (5.3.16)$$

where the back surface potential, $\Phi_{s2}(y)$, can be extracted from (5.3.9):

$$\Phi_{s2}(y)=\frac{C_{ox2}}{C_{ox2}+C_{si}}\left(V_{G2}-\Phi_{MS2}+\frac{Q_{ox2}}{C_{ox2}}+\frac{C_{si}}{C_{ox2}}\Phi_{s1}(y)+\frac{Q_{depl}}{2C_{ox2}}-\frac{Q_{s2}(y)}{C_{ox2}}\right) \qquad (5.3.17)$$

Different cases can now be distinguished for the back interface: the film can be fully depleted from source to drain (DS+DD = "depleted source + depleted drain"), accumulated from source to drain (AS+AD), accumulated near the source and depleted near the drain (AS+DD), inverted from source to drain (IS+ID), and inverted near the source and depleted near the drain (IS+DD). Because back-channel inversion is generally undesirable, we will describe only the (DS+DD), (AS+AD) and (AS+DD) cases [5.17].

* The back interface is accumulated from source to drain (AS+AD) under the condition that $V_{G2} < V_{G2,acc}(L)$

with
$$V_{G2,acc}(L) = V_{G2,acc} - \frac{C_{si}}{C_{ox2}} V_{DS} \tag{5.3.18.a}$$

where
$$V_{G2,acc} = \Phi_{MS2} - \frac{Q_{ox2}}{C_{ox2}} - \frac{C_{si}}{C_{ox2}} 2\Phi_F - \frac{Q_{depl}}{2C_{ox2}} \tag{5.3.18.b}$$

which is obtained from (5.3.9) and assuming $\Phi_{s1} = 2\Phi_F$, $\Phi_{s2} = 0$, and $Q_{s2}=0$. Equation (5.3.18.a) is also obtained from (5.3.9) with $\Phi_{s1}(L)= 2\Phi_F+V_{DS}$. Note that $V_{G2,acc}(L) < V_{G2,acc}(0)$.

From (5.3.15) and (5.3.16) one obtains:

$$I_{D,acc2} = \frac{W}{L} \mu_n C_{ox1} \left((V_{G1}-V_{th1,acc2})\, V_{DS} - \left(1 + \frac{C_{si}}{C_{ox1}}\right) \frac{V_{DS}^2}{2} \right) \tag{5.3.19}$$

where $V_{th1,acc2}$ is given by (5.3.10).

The drain saturation voltage is obtained from (5.3.19) under the condition that $dI_D/dV_{DS} = 0$:

$$V_{Dsat,acc2} = \frac{V_{G1}-V_{th1,acc2}}{1 + \frac{C_{si}}{C_{ox1}}} \tag{5.3.20}$$

and the saturation current is given by:

$$I_{Dsat,acc2} = \frac{1}{2} \frac{W}{L} \frac{\mu_n C_{ox1}}{1 + C_{si}/C_{ox1}} (V_{G1}-V_{th1,acc2})^2 \tag{5.3.21}$$

* The back interface is depleted from source to drain (DS+DD) under the condition that $V_{G2,acc} < V_{G2} < V_{G2,inv}$

where
$$V_{G2,inv} = \Phi_{MS2} - \frac{Q_{ox2}}{C_{ox2}} + 2\Phi_F - \frac{Q_{depl}}{2C_{ox2}} \tag{5.3.22}$$

which is obtained from (5.3.9) and assuming $\Phi_{s1} = \Phi_{s2} = 2\Phi_F$, and $Q_{s2}=0$.

The back interface being depleted from source to drain, we have $Q_{s2}(y)=0$. Using (5.3.15), (5.3.16) and (5.3.17), one obtains:

$$I_{D,depl2} = \frac{W}{L}\mu_n C_{ox1}\left((V_{G1}-V_{th1,depl2})\,V_{DS} - \left(1+\frac{C_{si}\,C_{ox2}}{C_{ox1}\,(C_{si}+C_{ox2})}\right)\frac{V_{DS}^2}{2}\right) \quad (5.3.23)$$

where the front threshold voltage under depleted back interface conditions, $V_{th1,depl2}$, is given by (5.3.12).

The drain saturation voltage is obtained from (5.3.23) under the condition that $dI_D/dV_{DS} = 0$:

$$V_{Dsat,depl2} = \frac{V_{G1}-V_{th1,depl2}}{1+\frac{C_{si}\,C_{ox2}}{C_{ox1}\,(C_{si}+C_{ox2})}} \quad (5.3.24)$$

and the saturation current is given by:

$$I_{Dsat,depl2} = \frac{1}{2}\frac{W}{L}\frac{\mu_n\,C_{ox1}}{1+\frac{C_{si}\,C_{ox2}}{C_{ox1}\,(C_{si}+C_{ox2})}}(V_{G1}-V_{th1,depl2})^2 \quad (5.3.25)$$

* If the back surface is accumulated near the source and depleted near the drain (AS+DD), the accumulation layer occurs at the back interface from y=0 to $y=y_t$, where y_t is the point where $\Phi_{s2}(y_t)=0$ and $Q_{s2}(y_t)=0$.

The drain current is then given by the following expression:

$$I_{D,AS+DD} = -\frac{W}{L}\mu_n\left(\int_{2\Phi_F}^{\Phi_{s1}(y_t)} Q_{inv1}(y)\,d\Phi_{s1}(y) + \int_{\Phi_{s1}(y_t)}^{2\Phi_F+V_{DS}} Q_{inv1}(y)\,d\Phi_{s1}(y)\right) \quad (5.3.26)$$

where $\Phi_{s1}(y_t)$ is given by combining (5.3.9) where $\Phi_{s2}(y_t)=0$ and $Q_{s2}(y_t)=0$ and (5.3.18.b):

$$\Phi_{s1}(y_t) = 2\Phi_F + \frac{C_{ox2}}{C_{si}}(V_{G2,acc} - V_{G2}) \quad (5.3.27)$$

Combining (5.3.26) and (5.3.27), one obtains:

$$I_{D,AS+DD} = \frac{W}{L} \mu_n C_{ox1} \left\{ (V_{G1} - V_{th1.acc2}) V_{DS} - \left(1 + \frac{C_{si}\, C_{ox2}}{C_{ox1}\,(C_{si}+C_{ox2})}\right) \frac{V_{DS}^2}{2} \right.$$
$$- \frac{C_{si}\, C_{ox2}}{C_{ox1}\,(C_{si}+C_{ox2})} V_{DS}\,(V_{G2,acc} - V_{G2})$$
$$\left. + \frac{C_{si}\, C_{ox2}}{C_{ox1}\,(C_{si}+C_{ox2})} \frac{C_{ox2}}{C_{si}} (V_{G2,acc} - V_{G2})^2 \right\} \qquad (5.3.28)$$

The drain saturation voltage is obtained from (5.3.23) under the condition that $dI_D/dV_{DS} = 0$:

$$V_{Dsat,AS+DD} = \frac{V_{G1} - V_{th1.acc2} - \frac{C_{si}\, C_{ox2}}{C_{ox1}\,(C_{si}+C_{ox2})}(V_{G2,acc} - V_{G2})}{1 - \frac{C_{si}\, C_{ox2}}{C_{ox1}\,(C_{si}+C_{ox2})}} \qquad (5.3.29)$$

and the drain saturation current is given by:

$$I_{Dsat,AS+DD} = \frac{W}{L} \frac{\mu_n C_{ox1}}{1 + \frac{C_{si} C_{ox2}}{C_{ox1}(C_{si}+C_{ox2})}} \left\{ (V_{G1} - V_{th1,acc2})^2 \right.$$
$$- \frac{2 C_{si} C_{ox2}}{C_{ox1}(C_{si}+C_{ox2})} (V_{G1} - V_{th1,acc2})(V_{G2,acc} - V_{G2})$$
$$\left. + \frac{C_{ox2}^2\,(C_{si}+C_{ox1})}{C_{ox1}^2\,(C_{si}+C_{ox2})} (V_{G2,acc} - V_{G2})^2 \right\} \qquad (5.3.30)$$

The saturation current in a thin-film SOI MOSFET (fully depleted with depleted back interface or fully depleted with accumulated back interface) is given by equations (5.3.21) and (5.3.25) and can be written in the general form:

$$I_{Dsat} \cong \frac{W\, \mu_n\, C_{ox1}}{2\, L\, (1 + \alpha)} [V_{G1} - V_{th}]^2 \qquad (5.3.31)$$

where $\alpha = C_{si}/C_{ox1}$ in a fully depleted device with accumulation at the back interface, and $\alpha = \frac{C_{si}\, C_{ox2}}{C_{ox1}\,(C_{si}+C_{ox2})}$ in a fully depleted transistor with depleted back interface. The AS+DD case basically follows similar rules, but it is more complicated and will not be described here. In a bulk transistor, expression (5.3.31) is also valid [5.30] if $\alpha = \frac{C_D}{C_{ox}}$ with C_D being equal to $\frac{\varepsilon_{si}}{x_{dmax}}$. Indeed, the Spice Level3 bulk MOSFET model [5.31] assumes $\alpha = \frac{\gamma}{2\sqrt{2\Phi f - V_{BS}}}$ with $\gamma = \frac{\sqrt{2\,\varepsilon_{si}\,q\,N_a}}{C_{ox}}$, which yields $\alpha = \frac{\varepsilon_{si}}{x_{dmax}\,C_{ox}}$ after straightforward manipulation. The values of the threshold voltages used in (5.3.31) can be extracted from relationships (5.3.10) and (5.3.12) for the SOI devices, and V_{th} has its usual meaning in the case of a bulk transistor. It is worth noting that the values of α are in the following sequence:

$$\alpha_{\text{fully depleted SOI}} < \alpha_{\text{bulk}} < \alpha_{\text{back accum SOI}}$$

so that the drain saturation current is highest in the fully depleted device, lower in the bulk device, and even lower in the device with back accumulation. It is also interesting to note that α represents the ratio C_b/C_{ox1} of two capacitors. Indeed, C_{ox1} is the gate oxide capacitance, and C_b is the capacitance between the inversion channel and the back-gate electrode. The value of C_b is either that of the depletion capacitance $C_D=\varepsilon_{si}/x_{dmax}$ in a bulk device, the silicon film capacitance $Csi=\varepsilon_{si}/t_{si}$ in a fully depleted device with back interface accumulation, or the value given by the series association of C_{si} and C_{ox2} ($C_b = \frac{C_{si}C_{ox2}}{(C_{si}+C_{ox2})}$) in a fully depleted device.

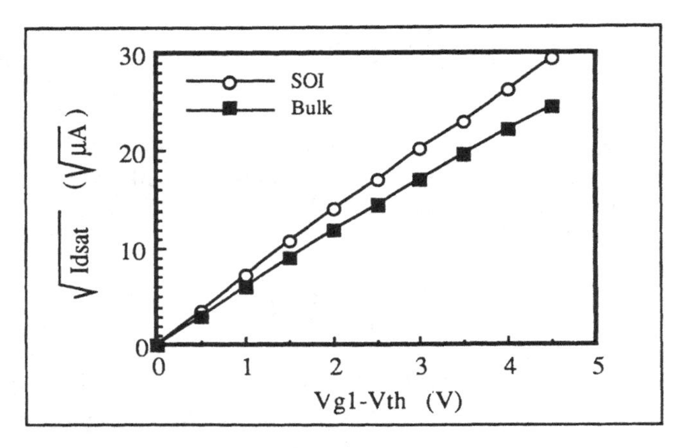

Figure 5.3.9: $\sqrt{I_{Dsat}}$ as a function of $V_{G1}-V_{th}$ in a bulk and a thin-film, fully depleted SOI device with same technological parameters.

This high saturation current in thin-film, fully depleted SOI MOSFETs brings about an increase of current drive (compared to bulk devices) which largely contributes to the excellent speed performances of thin-film fully depleted (FD) SOI CMOS circuits. Figure 5.3.9 compares the drain saturation current in a bulk and a thin-film FD SOI device, as a function of $V_{G(1)}-V_{th}$. The larger current drive of the SOI device is evident [5.29] (it is actually 20-30% higher that the drive of the bulk device).

It is worth noting that the SOI MOSFET, which was considered so far as a four-terminal device (source, drain, front gate and back gate), is actually a five-terminal device, the fifth terminal being the body. In many applications, the body is left uncontacted, but its presence has considerable influence on the dynamic characteristics of the device. This influence is more pronounced in thick-film devices than in thin, fully depleted devices. A charge-based model of the five-terminal SOI MOSFET can be found in Ref. [5.18]. Detailed analysis of the short-channel effects is described in Ref. [5.10].

5.4. Transconductance and mobility

The current drive capability of an SOI MOSFET depends basically on three factors: the saturation current, the transconductance and the mobility of the carriers in the inversion layer. The saturation current has been dealt with in Section 5.3, and it has been shown that I_{Dsat} is larger in a thin-film, fully depleted SOI MOSFET with depleted back interface than in the corresponding bulk device. On the other hand, I_{Dsat} is lower in a thin-film, fully depleted SOI MOSFET with accumulated back interface than in a bulk device. The transconductance and the channel mobility are described next.

5.4.1 Transconductance

The transconductance of a MOSFET, g_m, is a measure of the effectiveness of the control of the drain current by the gate voltage. In a bulk n-channel MOSFET in saturation, it is given by:

$$g_m = dI_{Dsat}/dV_G \qquad (\text{for } V_{DS} > V_{Dsat})$$

$$= \frac{W}{L}\mu_n C_{ox}\left(V_G-V_{th}\right) + 4\Phi_F\left(\frac{C_D}{C_{ox}}\right)^2\left(1-\sqrt{1+\left(\frac{C_{ox}}{C_{depl}}\right)^2\frac{V_G-V_{FB}}{2\Phi_F}}\right)$$

with $C_D=\varepsilon_{si}/x_{dmax}$ [5.52]

which can be approximated by (from 5.3.31):

$$I_{Dsat} \cong \frac{W\ \mu_n\ C_{ox1}}{2\ L\ (1\ +\ \alpha)}(V_{G1}\text{-}V_{th})\ \text{with}\ \ \alpha = \frac{\varepsilon_{si}}{x_{dmax}\ C_{ox}}$$

In a thin-film, fully-depleted SOI MOSFET, the transconductance can be obtained from (5.3.21) and (5.3.25):

$$g_m = dI_{Dsat}/dV_{G1} = \frac{W\ \mu_n\ C_{ox1}}{2\ L\ (1\ +\ \alpha)}(V_{G1}\text{-}V_{th}) \qquad (5.4.1)$$

for $V_{DS} > V_{Dsat}$ and with $\alpha = C_{si}/C_{ox1}$ in a fully depleted device with accumulation at the back interface, and $\alpha = \frac{C_{si}\ C_{ox2}}{C_{ox1}\ (C_{si}+C_{ox2})}$ in a fully depleted transistor with depleted back interface [5.29].

As in the case of the analysis of the saturation current (Section 5.3), α represents the ratio C_b/C_{ox1} of two capacitors. Indeed, C_{ox1} is the gate oxide capacitance, and C_b is the capacitance between the inversion channel and the back-gate electrode. The value of C_b is either that of the depletion capacitance $C_D=\varepsilon_{si}/x_{dmax}$ in a bulk device, the silicon film capacitance $Csi=\varepsilon_{si}/t_{si}$ in a fully depleted device with back interface accumulation, or the value given by the series association of C_{si} and C_{ox2} ($C_b = \frac{C_{si}C_{ox2}}{(C_{si}+C_{ox2})}$) in a fully depleted device. It is once again worth noting that the values of α are typically in the following sequence:

$$\alpha_{\text{fully depleted SOI}} < \alpha_{\text{bulk}} < \alpha_{\text{back accum SOI}}$$

such that the transconductance is highest in the fully depleted device, lower in the bulk device, and even lower in the device with back accumulation.

5.4.2. Mobility

The mobility of the electrons within the inversion layer of an n-channel MOSFET has been so far considered as a constant. It is actually a function of the vertical electric field below the gate oxide. It can be approximated by:

$$\mu_n(y) = \mu_{max}\ [E_c/E_{eff}(y)]^c \ \ \text{for}\ E_{eff}(y) > E_c \qquad (5.4.2)$$

where μ_{max}, E_c, and c are fitting parameters depending on the gate oxidation process and the device properties [5.17, 5.24], and

$$E_{eff}(y) = E_{s1}(y) - \frac{Q_{inv1}(y)}{2\varepsilon_{si}} \qquad (5.4.3)$$

The vertical electric field below the gate oxide is given by (5.3.4):

$$E_{s1}(y) = \left(\frac{\Phi_{s1}(y)-\Phi_{s2}(y)}{t_{si}} + \frac{q\ N_a\ t_{si}}{2\ \varepsilon_{si}}\right)$$

The inversion charge in the channel, $Q_{inv1}(y)$, can be obtained from (5.3.16):

$$-Q_{inv1}(y)=C_{ox1}\left(V_{G1}-\Phi_{MS1}+\frac{Q_{ox1}}{C_{ox1}} - \left(1+\frac{C_{si}}{C_{ox1}}\right)\Phi_{s1}(y)+\frac{C_{si}}{C_{ox1}}\Phi_{s2}(y) + \frac{Q_{depl}}{2C_{ox1}}\right)$$

and the back surface potential, $\Phi_{s2}(y)$, is given by (5.3.17):

$$\Phi_{s2}(y)= \frac{C_{ox2}}{C_{ox2}+C_{si}}\left(V_{G2}-\Phi_{MS2}+\frac{Q_{ox2}}{C_{ox2}} + \frac{C_{si}}{C_{ox2}}\Phi_{s1}(y) + \frac{Q_{depl}}{2C_{ox2}} - \frac{Q_{s2}(y)}{C_{ox2}}\right)$$

The expression of the surface electric field can be simplified by considering the case of a thin, fully depleted devices operating with a low drain voltage ($V_{DS}\cong0$), so that the front and back surface potentials are independent of *y*. If the back interface is depleted and close to inversion, $\Phi_{s1}-\Phi_{s2} \cong 0$, and the front surface electric field is equal to $\frac{q\ N_a\ t_{si}}{2\ \varepsilon_{si}}$. In this case, the surface electric field is lower than the surface electric field in the corresponding bulk device, which is equal to $\frac{qN_a x_{dmax}}{\varepsilon_{si}}$, since $t_{si} < x_{dmax}$. If the film is fully depleted, but if the film is not close to inversion, $E_{s1} \cong \frac{q\ N_a\ x_1}{\varepsilon_{si}}$ is a good approximation (x_1 is the point of minimum potential in the film - Figures 5.3.2 and 5.4.1), and the surface electric field is still lower than in a bulk devices, since $x_1 < t_{si} < x_{dmax}$.

Figure 5.4.1 presents the electric field as a function of depth, *x*, in a bulk and a thin-film, fully depleted SOI device. The slope of the E(x) lines are the same if both devices have the same doping concentration, N_a. The (front) surface electric field, E_{s1}=E(x=0), is lower in the SOI device than in the bulk device. It is worth noting, however, that if the back of the fully depleted SOI device is accumulated ($\Phi_{s1}-\Phi_{s2} \cong 2\Phi_F$), the surface electric field can be larger than in the corresponding bulk device.

The increase of surface mobility in thin, fully depleted SOI devices has been reported by several authors [5.25, 5.26]. The maximum of mobility is obtained right above threshold, but is rapidly overshadowed by the decrease of mobility caused by the increase of Q_{inv1} when V_{G1} is significantly larger than V_{th1} (see equation (5.4.3)).

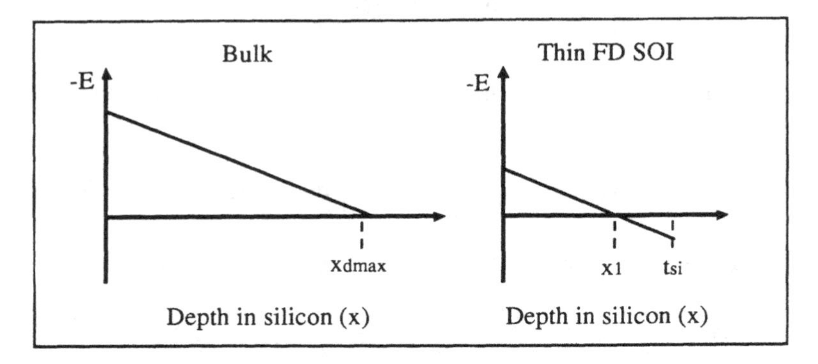

Figure 5.4.1: Electric field distribution in a bulk and a thin-film, fully depleted (FD) SOI device with depleted back interface.

Beyond equation (5.4.3), the reduction of the surface electric field means a broadening of the inversion channel, which leads to reduced scattering of the carriers in the inversion channel by the Si/SiO_2 interface. In an extreme case, the whole thickness of the silicon film can become inverted. In that case, the inversion carriers located near the center of the film (*i.e.* at some distance from the interfaces) exhibit bulk-like mobility (in contrast to surface mobility) [5.27]. The complete description of thin-film devices with volume inversion implies quantum-mechanical considerations which lie beyond the scope of the present analysis.

5.5. Subthreshold slope

The inverse subthreshold slope (or, in short, the subthreshold slope, or subthreshold swing) is defined as the inverse of the slope of the Id(Vg) curve in the subthreshold regime, presented on a semilogarithmic plot (Figure 5.5.1).

$$S = \frac{d\ V_G}{d\ (\log I_D)} \qquad (5.5.1)$$

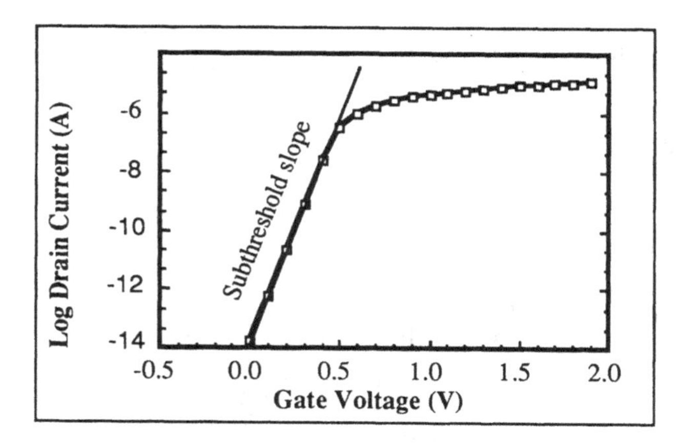

Figure 5.5.1: Semilogarithmic plot of $I_D(V_G)$ of an nMOS device.

The subthreshold current of an MOS transistor is a minority carrier diffusion current [5.19]. With analogy to bipolar transistor theory (current in the base), we find:

$$I_D = -q\ A\ D_n \frac{dn}{dy} = q\ A\ D_n \frac{n(0)-n(L)}{L} \qquad (5.5.2)$$

with A being the cross section of the channel area, D_n is the diffusion coefficient of electrons, and n(0) and n(L) the electron (minority carrier) concentrations at the source and drain side. We have: $n(0)=n_{po}\ e^{\beta\Phi_s}$ and $n(L)=n_{po}\ e^{\beta(\Phi_s - V_D)}$ where Φ_s is the surface potential $\beta=\frac{q}{kT}$, and $n_{po}=\frac{n_i^2}{N_a}$. The inversion channel can be approximated by a box carrier profile with a uniform carrier concentration n(0) and a thickness *d* defined as the distance from the $Si\text{-}SiO_2$ interface at which the potential is lowered by $\frac{kT}{q}$ [5.20].

We have: $$d = \frac{\frac{kT}{q}}{E_s} \quad \text{with } E_s = -\frac{d\Phi_s}{dx}.$$

Using Einstein's relationship $D_n = \frac{kT}{q}\mu_n$ and $A=W.d$, where W is the channel width, we find:

$$I_D = \left(\mu_n \frac{W}{L} q \left(\frac{kT}{q}\right)^2 \frac{n_i^2}{N_a}(1 - e^{-\beta V_D})\right) \frac{e^{\beta\Phi_s}}{-\frac{d\Phi_s}{dx}} \qquad (5.5.3)$$

The inverse subthreshold slope S is given by $\frac{dV_G}{d\log(I_D)}$

and is equal to $S = \dfrac{\ln(10)}{\dfrac{d \ln I_D}{d V_G}}$ (5.5.4)

From eqn (5.5.3) we derive:

$$\frac{d\ln I_D}{dV_G} = \frac{1}{I_D}\frac{dI_D}{d\Phi_s}\frac{d\Phi_s}{dV_G} = C\frac{d\Phi_s}{dV_G} \tag{5.5.5}$$

with

$$C = \frac{q}{kT} - \frac{\dfrac{d}{d\Phi_s}\left(-\dfrac{d\Phi_s}{dx}\right)}{-\dfrac{d\Phi_s}{dx}} \tag{5.5.6}$$

The second term of C is a correction term which accounts for the reduction of current increase due to the increase of surface electric field with increasing surface current (a large surface electric field will decrease the channel thickness).

In the case of a **bulk** or a **thick** SOI device, we have:

$E_s = -\dfrac{d\Phi_s}{dx} = \sqrt{\dfrac{2\,q\,N_a\,\Phi_s}{\varepsilon_{si}}} = \dfrac{q\,N_a}{C_D}$ with C_D being the depletion capacitance. We also have: $\dfrac{d}{d\Phi_s}\left(-\dfrac{d\Phi_s}{dx}\right) = \dfrac{C_D}{\varepsilon_{si}}$. The correction term is then given by:

$$\frac{\dfrac{d}{d\Phi_s}\left(-\dfrac{d\Phi_s}{dx}\right)}{-\dfrac{d\Phi_s}{dx}} = \frac{C_D^2}{q\,N_a\,\varepsilon_{si}} = \frac{1}{2\Phi_s} \tag{5.5.7}$$

Using the relationship between gate voltage and charges in the silicon and at the interfaces:

$$V_G = \Phi_{MS} + \Phi_s + \frac{-Q_D - Q_{ox} + C_{it}\,\Phi_s}{C_{ox}}$$

we find:

$$\frac{d\Phi_s}{dV_G} = \frac{C_{ox}}{C_{ox}+C_D+C_{it}} \tag{5.5.8}$$

where $C_D = dQ_D/d\Phi_S$, $Q_D = q\,N_a\,x_{dmax}$, and $C_{it} = q\,N_{it}$.

Usually, the correction factor is small compared to $\frac{q}{kT}$ (about 4% of it, considering $\Phi_s = \frac{2}{3}\Phi_F$), and will be neglected here. Using (5.5.4), (5.5.5) and (5.5.6), one finds:

$$S=\frac{kT}{q}\ln(10)\left(1+\frac{C_D+C_{it}}{C_{ox}}\right) \quad (5.5.9)$$

or

$$S=\frac{kT}{q}\ln(10)\left(1+\frac{C_D}{C_{ox}}\right) \quad (5.5.10)$$

if the interface traps are neglected.

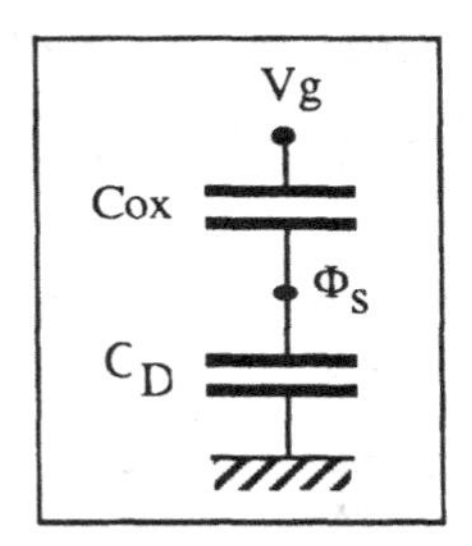

Figure 5.5.2: Equivalent capacitor network (bulk case).

The expression of the inverse subthreshold slope is the same for a thick-film (partially depleted) SOI device. In both cases, the right-side term of (5.5.10) can be represented by the capacitor network of Figure 5.5.2 in which we can observe that

$$C_{ox}\, d(V_G-\Phi_s) = Ig = C_D\, d\,\Phi_s \text{ and, hence, } \frac{d\,V_G}{d\,\Phi_s} = 1 + \frac{C_D}{C_{ox}}.$$

In the case of a thin-film, **fully depleted** SOI device (with depleted back interface), we have from equations (5.3.8) and (5.3.9):

$$V_{G1} = \Phi_{MS1} - \frac{Q_{ox1}}{C_{ox1}} + \frac{q\,N_a\,t_{si}}{2\,C_{ox1}} + \Phi_{s1}\left(\frac{\varepsilon_{si}}{t_{si}\,C_{ox1}}+1\right) + \Phi_{s2}\left(\frac{-\varepsilon_{si}}{t_{si}\,C_{ox1}}\right) \quad (5.5.11)$$

$$V_{G2} = \Phi_{MS2} - \frac{Q_{ox2}}{C_{ox2}} + \frac{q\,N_a\,t_{si}}{2\,C_{ox2}} + \Phi_{s1}\left(\frac{-\varepsilon_{si}}{t_{si}\,C_{ox2}}\right) + \Phi_{s2}\left(\frac{\varepsilon_{si}}{t_{si}\,C_{ox2}}+1\right) \quad (5.5.12)$$

Writing $\varepsilon_{si}/t_{si}=C_{si}$, the capacitance of the depleted silicon film, and $q\,N_a\,t_{si} = Q_{depl}$, the depletion charge the silicon film, we obtain:

$$V_{G1} = \Phi_{MS1} - \frac{Q_{ox1}}{C_{ox1}} + \frac{Q_{depl}}{2\,C_{ox1}} + \Phi_{s1}(\frac{C_{si}}{C_{ox1}} + 1) + \Phi_{s2}(\frac{-C_{si}}{C_{ox1}}) \qquad (5.5.13)$$

$$V_{G2} = \Phi_{MS2} - \frac{Q_{ox2}}{C_{ox2}} + \frac{Q_{depl}}{2\,C_{ox2}} + \Phi_{s1}(\frac{-C_{si}}{C_{ox2}}) + \Phi_{s2}(\frac{C_{si}}{C_{ox2}} + 1) \qquad (5.5.14)$$

Eliminating Φ_{s2} between these two equations, we can obtain Φ_{s1} as a function of V_{G1} and V_{G2}:

$$\Phi_{s1}\left((1 + \frac{C_{si}}{C_{ox1}})\frac{C_{ox1}}{C_{si}}(1 + \frac{C_{si}}{C_{ox2}}) - \frac{C_{si}}{C_{ox2}}\right) = \qquad (5.5.15)$$

$$V_{G2} - \Phi_{MS2} + \frac{Q_{ox2}}{C_{ox2}} - \left(\frac{Q_{depl}}{2\,C_{ox1}} + \Phi_{MS1} - \frac{Q_{ox1}}{C_{ox1}} - V_{G1}\right)\frac{C_{ox1}}{C_{si}}(1 + \frac{C_{si}}{C_{ox2}}) + \frac{Q_{depl}}{2\,C_{ox2}}$$

Assuming back-channel conduction is negligible [5.20], we have to calculate the value of $\frac{d\ln I_D}{dV_{G1}} = \frac{1}{I_D}\frac{dI_D}{d\Phi_{s1}}\frac{d\Phi_{s1}}{dV_{G1}}$

From equation (5.5.15) we obtain:

$$\frac{d\Phi_{s1}}{dV_{G1}} = \frac{\frac{C_{ox1}}{C_{si}}(1 + \frac{C_{si}}{C_{ox2}})}{(1 + \frac{C_{si}}{C_{ox1}})\frac{C_{ox1}}{C_{si}}(1 + \frac{C_{si}}{C_{ox2}}) - \frac{C_{si}}{C_{ox2}}} = \frac{\frac{1}{C_{si}} + \frac{1}{C_{ox2}}}{\frac{1}{C_{ox1}} + \frac{1}{C_{si}} + \frac{1}{C_{ox2}}} \qquad (5.5.16)$$

This expression corresponds to the capacitor network of Figure 5.5.3.

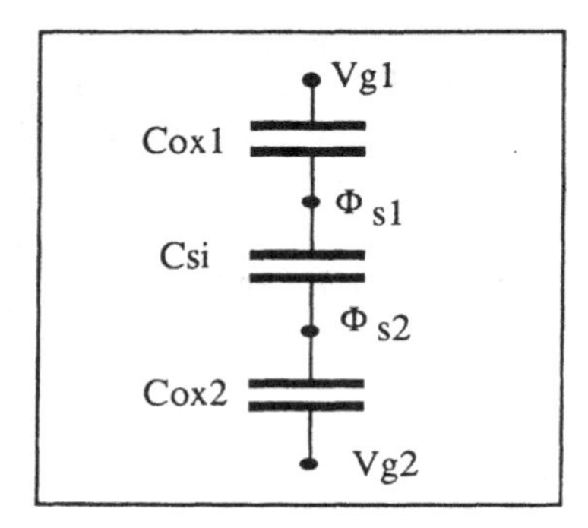

Figure 5.5.3: Equivalent capacitor network (thin-film, fully depleted SOI case).

It can be shown that the correction factor, C, is smaller in thin SOI devices than in bulk devices, and can be neglected.

Usually, $C_{ox2} << C_{ox1}$ and $C_{ox2} << C_{si}$. As a result, $\frac{d\Phi_{s1}}{dV_{G1}}$ tends to unity, and, using (5.5.4), (5.5.5) and (5.5.6) one obtains:

$$S \cong \frac{kT}{q} \ln (10) \tag{5.5.17}$$

A mere comparison of equations (5.5.10) and (5.5.17) shows that the inverse subthreshold slope of a thin-film, fully depleted SOI MOSFET will be lower than that of a bulk or thick-film SOI device having the same parameters. The theoretical minimum value of S is 60 mV/decade at room temperature (a 60 mV increase of gate voltage results into a tenfold increase of subthreshold drain current). This also means that, in the subthreshold regime, any increase of gate bias ΔV_G will give rise to an increase of surface potential $\Delta\Phi_{s1}$ equal to ΔV_G (perfect coupling between V_G and Φ_{s1}).

A comparison between the subthreshold slopes in a thin-film SOI MOSFET and a thicker-film device is made in Figure 5.5.4, where the $I_D(V_G)$ characteristics of two n-channel devices (a thin-film device with t_{si}=100 nm and a device exhibiting thick-film behavior (t_{si}=200 nm) are presented [5.12]. The subthreshold slope of the thicker-film device is identical to that of a bulk device with same channel dopant concentration. It is interesting to note that the thinner device has a slightly lower leakage current at V_G=0V than the thicker device, although its threshold voltage is lower.

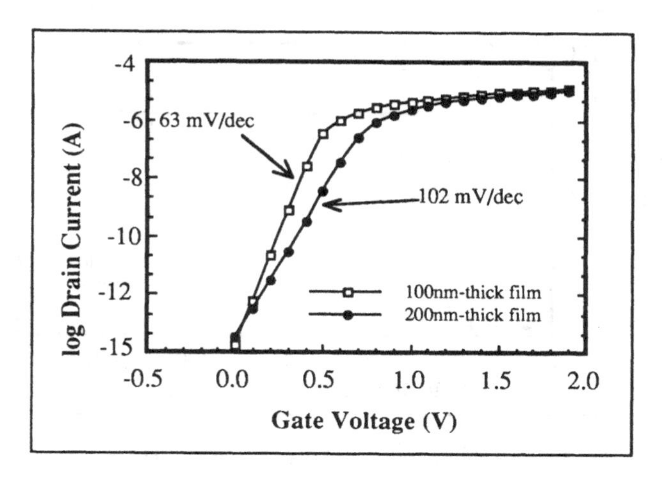

Figure 5.5.4: Simulated $I_D(V_G)$ characteristics of a 200 nm-thick, partially depleted n-channel device and of a 100 nm-thick, fully depleted device MOSFET.

In general, and neglecting the presence of interface states, one can write:

$$S = \frac{kT}{q} \ln (10)\ (1+ \alpha) \qquad (5.5.18)$$

where, as in the case of the analyses of the saturation current (Section 5.3) and the transconductance (Section 5.4), α represents the ratio C_b/C_{ox1} of two capacitors. Indeed, C_{ox1} is the gate oxide capacitance, and C_b is the capacitance between the inversion channel and the back-gate electrode. The value of C_b is either that of the depletion capacitance $C_D=\varepsilon_{si}/x_{dmax}$ in a bulk device, the silicon film capacitance $Csi=\varepsilon_{si}/t_{si}$ in a fully depleted device with back interface accumulation, or the value given by the series association of C_{si} and C_{ox2} ($C_b = \frac{C_{si}C_{ox2}}{(C_{si}+C_{ox2})}$) in a fully depleted device. It is once again worth noting that the values of α are typically in the following sequence:

$$\alpha_{\text{fully depleted SOI}} < \alpha_{\text{bulk}} < \alpha_{\text{back accum SOI}}$$

such that the inverse subthreshold slope has the lowest (*i.e.* best) value in the fully depleted device, it is larger in the bulk device, and even larger in the device with back accumulation.

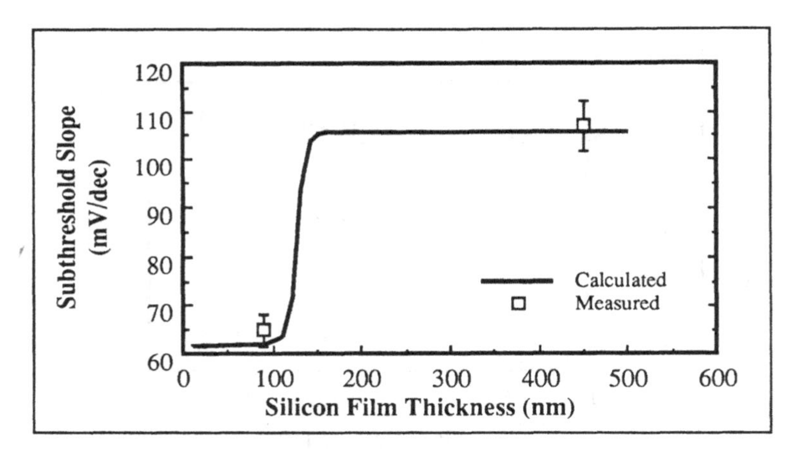

Figure 5.5.5: Simulated and measured subthreshold slope as a function of silicon film thickness. N_a = 8 10^{16} cm^{-3}, gate oxide thickness is 25 nm [5.21].

The minimum theoretical theoretical value of 60 mV/dec at room temperature is never reached because of the presence of traps at the Si-SiO2 interfaces and because the finite value of C_{ox2} (Figure 5.5.3), but values as low as 65 mV/dec can easily be obtained. The transition between the thin-film and the thick-film regimes is quite sharp, as illustrated in Figure 5.5.5, where the transition between partial and full depletion is quite clear.

The excellent value of the subthreshold slope in thin-film, fully-depleted SOI devices allows one to use smaller values of threshold voltage than in bulk (or thick SOI) devices without increasing the leakage current at V_{G1}=0. As a result, better speed performances can be obtained, especially at low supply voltages (2-3 volts). [5.22] As in bulk devices, an increase of the subthreshold slope is observed in short-channel devices, but thin SOI MOSFETs show less degradation than bulk transistors [5.22][5.23].

5.6. Impact ionization and high-field effects

Several parasitic phenomena related to impact ionization in the high electric field region near the drain appear in SOI MOSFETs. Some of them, such as the reduced drain breakdown voltage, are related to the parasitic NPN bipolar transistor found in the n-channel SOI MOSFET, and will be dealt with in Section 5.7. The present section will neglect body current multiplication by bipolar effect and focus on two phenomena. The first of these is the kink effect, and the second one is the hot-electron degradation. The kink effect is normally not found in bulk devices operating at room temperature when substrate or well contacts are provided, which is usually the case. Hot-electron degradation takes place in bulk MOSFETs having a short channel and constitutes a major reliability hazard in submicron devices. A comparison of hot-electron degradation phenomena between bulk and SOI devices will be made in Section 5.6.2.

5.6.1. Kink effect

The kink effect consists into the appearance of a "kink" in the output characteristics of an SOI MOSFET, as illustrated in Figure 5.6.1. The kink appears above a given drain voltage. It can be very strong in n-channel transistors, but is usually absent from p-channel devices. The kink effect is not observed in bulk devices at room temperature when substrate or well contacts are provided, but can be observed in bulk MOSFETs operating at low temperature [5.32] or in devices realized in floating (uncontacted) p-wells. Finally, it has been observed from the very beginning of SOI technology, namely in SOS devices [5.33].

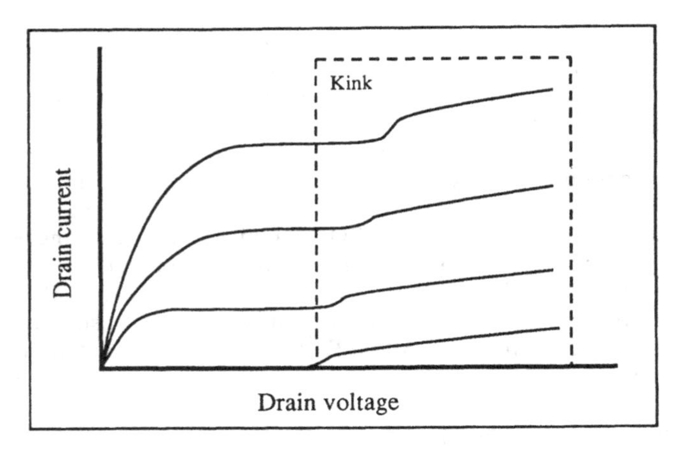

Figure 5.6.1: Illustration of the kink effect in the output characteristics of an n-channel SOI MOSFET.

The kink effect can be explained as follows. Let us consider a "thick"-film, **partially depleted** SOI n-channel transistor. When the drain voltage is high enough, the channel electrons can acquire sufficient energy in the high electric field zone near the drain to create electron-hole pairs, due to an impact ionization mechanism. The generated electrons rapidly move into the channel and the drain, while the holes (which are majority carriers in the p-type body) migrate towards the place of lowest potential, *i.e.*, the floating body (Figure 5.6.2.C).

The injection of holes into the floating body forward biases the source-body diode. The floating body reaches a positive potential which can be calculated by writing the following equation:

$$I_{holes,gen} = I_{so}\left(\exp\left(\frac{q\,V_{BS}}{n\,k\,T}\right) - 1\right)$$

where $I_{holes,\,gen}$ is the hole current generated near the drain (Figure 5.6.2.A), I_{so} is the saturation current of the source-body diode, V_{BS} is the potential of the floating body, and *n* is the ideality factor of the diode. The value of $I_{holes,gen}$ depends on different parameters, including the body potential. Its calculation requires iterative solving of a set of complex equations [5.34] and will not be described here.

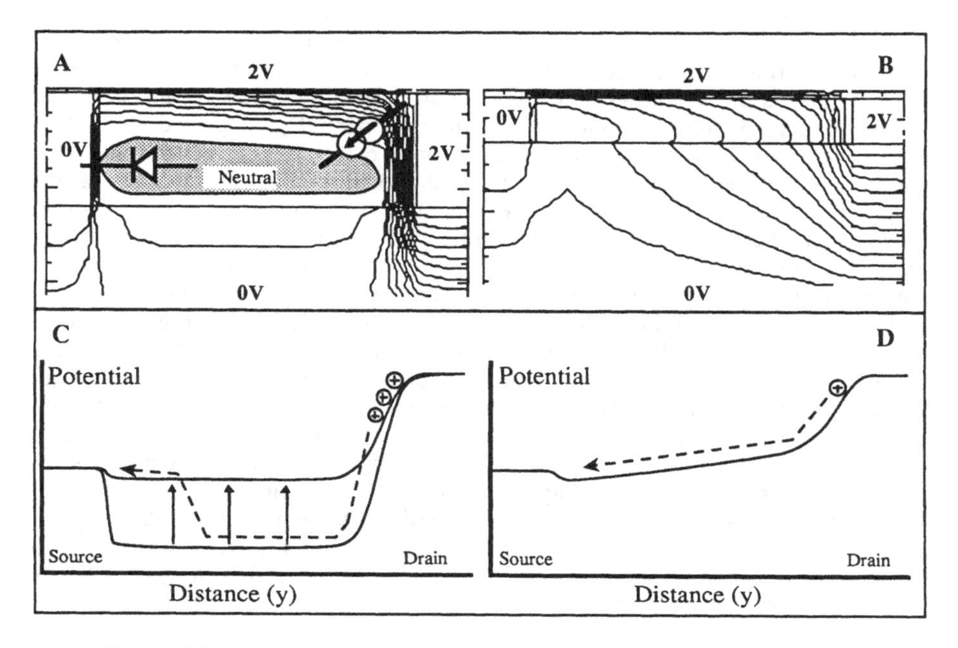

Figure 5.6.2:
A: Isopotential lines in a "thick"-film (30 nm, $N_a=8 \times 10^{16}$ cm^{-3}), partially depleted (PD) SOI n-channel MOSFET (one curve every 200 mV). The grey area is the neutral, floating body.
B: Isopotential lines in a thin-film (10 nm, $N_a=8 \times 10^{16}$ cm^{-3}), fully depleted (FD) device.
C: Potential in the neutral region from source to drain in the PD device before and after the onset of the kink effect (lower and upper curve, respectively).
D: Potential from source to drain in the FD device.

The increase of body potential (an increase of 700-800 mV) gives rise to a decrease of the threshold voltage. This is again a manifestation of the body effect (see Section 5.3.2). The threshold voltage shift can be calculated using the expression of the body effect in a ***bulk*** transistor where the substrate bias is replaced by the floating body potential. This decrease of threshold voltage induces an increase of the drain current as a function of drain voltage which can be observed in the output characteristics of the device (Figure 5.6.1), and which is called "kink effect". If the minority carrier lifetime in the silicon film is high enough, the kink effect can be reinforced by the NPN bipolar transistor structure present in the device (the "base" hole current is amplified by the bipolar gain, which gives rise to an increased net drain current, sometimes called "second kink") [5.33].

Let us now consider the case of a thin-film, **fully depleted** SOI n-channel transistor (Figure 5.6.2.B). A mere comparison of Figures 5.6.2.A and 5.6.2.B shows that the electric field near the drain is lower in the fully depleted device than in the partially depleted one (the density of isopotential lines is lower). As a result, less electron-hole pair generation will take place in the fully depleted device [5.35]. As in the case of the partially depleted device, the generated electrons rapidly move into the channel and the drain, while the holes migrate towards the place of lowest potential, *i.e.*, near the source junction. However, contrarily to the case of the partially depleted transistor, the source-to-body diode is "already forward biased" (the body-source potential barrier is very small), due to the full depletion of the film (Figure 5.6.2.D), and the holes can readily recombine in the source without having to raise the body potential (there is no significant potential barrier between the body and the source). As a result, the body potential remains unchanged, the body effect is virtually equal to zero, and there is no threshold voltage decrease as a function of drain voltage. This explains why thin-film, fully depleted n-channel SOI MOSFETs are free of kink effect. If a negative back-gate bias is used, however, to induce an accumulation layer at the back interface, the device behaves as a partially depleted device, and the kink reappears [5.35].

For completeness, one can mention that the p-channel SOI transistors are usually free of kink effect because coefficient of pair generation by energetic holes is much lower than that of pair generation by energetic electrons. The kink effect is not observed in bulk devices if the majority carriers generated by impact isolation can escape into the substrate or to a well contact. If the well is left floating, or if the silicon is not conducting (*e.g.* at such low temperatures that the carriers are frozen out), the kink effect appears. Finally, the kink effect can be eliminated from partially depleted SOI MOSFETs if a substrate contact is provided for the removal of excess majority carriers from the device body (this technique is unfortunately not 100% effective because of the relatively high resistance of the body) (see also Section 5.7.1).

The presence of a floating substrate can also give rise to anomalous subthreshold slope effects [5.53]. Indeed, when the body floats, the (relatively weak) impact ionization which can occur near the drain in the subthreshold regime (in addition to the other carrier generation mechanisms) can result in holes being injected into the neutral body. The hole injection charges the body and produces a forward bias on the body-source junction which reduces the threshold voltage while the device operates in the subthreshold region. As a result, the subthreshold current "jumps" from a high-V_{th} characteristics to a low-V_{th} $I_D(V_{G1})$ curve, and a subthreshold slope lower than 60 mV/decade (at room temperature) can be observed [5.53].

5.6.2. Hot-electron degradation

Hot-carrier degradation is one of the most important factors affecting the reliability of modern MOS devices. During the last decade, the dimensions of the devices have been dramatically reduced, but the supply voltage of integrated circuits has remained fixed at a steady 5 volts. Submicron devices would need a lower supply voltage, such as 3.3 volts. Such a decrease of supply voltage, however, would reduce the speed performance of the circuits, and the pressure is high to keep compatibility with the 5V standard as long as possible.

The horizontal electric field in the transistor, which is roughly proportional to the ratio of the supply voltage to the gate length, increases as the device dimensions are reduced, and, in modern MOSFETs, the electric field near the drain can reach values high enough to affect the reliability of the device. Let us take the example of an n-channel device. When the transistor operates in the saturation mode, a high electric field can develop between the channel pinchoff and the drain junction. This electric field gives the electrons such a high energy that some of them can be injected into the gate oxide, thereby damaging the oxide-silicon interface [5.36]. At very high injection levels, gate current can even be measured [5.37]. The high-energy channel electrons can also create electron-hole pairs by impact ionization. In a bulk device, the generated holes escape into the substrate and constitute a substrate current. There exists a relationship between the substrate current and the gate current [5.38], and the lifetime of the device can be related to the magnitude of the hot-electron injection on the gate oxide. The relation between the device lifetime, τ, defined by (static) hot-carrier degradation of the gate oxide, can be correlated with the impact-ionization current according to the following relationship [5.30]:

$$\tau \propto \frac{W}{I_D} (M-1)^{-m} \qquad (5.6.1)$$

where $m \cong 3$ [5.39], and M is the multiplication factor due to impact ionization, defined by $I_{body} = (M-1)\, I_{Dsat}$, where I_{body} is the hole current generated by impact ionization. In a fully depleted SOI MOSFET, the multiplication factor can be obtained by integrating the ionization coefficient over the high field region near the drain, and depends on both the drain and gate voltages [5.30]:

$$M-1 \cong \frac{A_i}{B_i} [V_{DS}-V_{Dsat}] \exp\left(-\frac{B_i\, l_c}{V_{DS}-V_{Dsat}}\right)$$

where l_c is a characteristic length defined by: $l_c = t_{si}\left(\frac{C_{si}\,\beta}{2C_{ox1}\,(1+\alpha)}\right)$ with $\alpha = 1$ if the back interface is accumulated, and $\alpha = 1 + \frac{C_{si}}{C_{si}+C_{ox2}}$ if the back interface is depleted. A_i and B_i are impact ionization constants for electrons, assumed to be equal to 1.4×10^6 cm^{-1} and 2.6×10^6 V/cm, respectively.

Figure 5.6.3 presents the multiplication factor (M-1) in bulk and thin-film, fully depleted SOI MOSFETs of similar geometries [5.30]. The multiplication factor, is related to the lifetime of the device through expression (5.6.1). Clearly, less carrier-electron degradation can be expected from the fully depleted SOI MOSFETs. Numerical modelling of the devices indeed shows that the electric field near the drain is minimized if fully-depleted SOI devices are used [5.40].

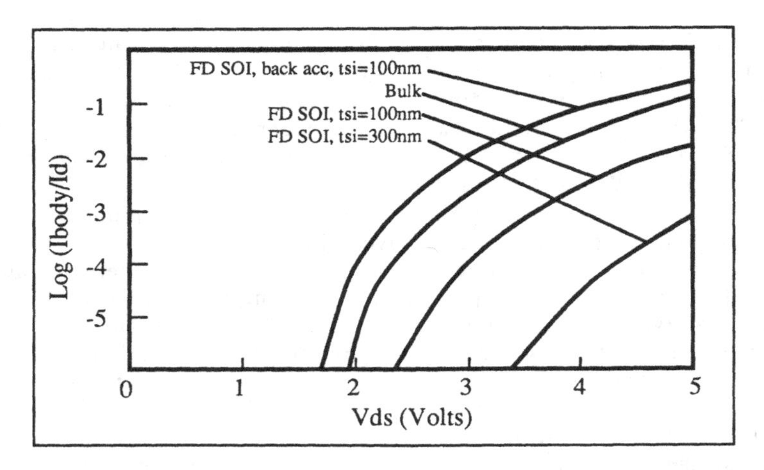

Figure 5.6.3: Multiplication factor $(M-1)=I_{body}/I_D$ as function of drain voltage in bulk and fully-depleted SOI nMOSFETs. The SOI MOSFETs have a thickness of either 100 or 300 nm, with or without back accumulation. L=1μm, V_G-V_{th}=1.5 V [5.30].

It is worth noting that devices with back accumulation can present higher drain electric fields than bulk devices and can, therefore, be submitted to more hot-carrier degradation [5.30]. Finally, the use of thicker SOI films permits one to further minimize hot-electron degradation, provided the device remains fully depleted [5.30]. Confirmation of the reduced hot-electron degradation in fully depleted SOI devices has been experimentally obtained. Both threshold voltage shift and transconductance degradation as a function of stress time at high V_{DS} have been found to be significantly smaller in fully depleted devices than in bulk devices and SOI devices with back-interface accumulation [5.40].

5.7. Parasitic bipolar effects

There exists a parasitic bipolar transistor in the MOS structure. If we consider an n-channel device, the N^+ source, the P-type body and the N^+ drain indeed form the emitter, the base, and the collector of an NPN bipolar transistor. In a bulk device, the base of the bipolar transistor is usually grounded by means of a substrate contact. In an SOI device, however, the body (= the base of the bipolar transistor) is usually left floating. This parasitic bipolar transistor is the origin of several undesirable effects in SOI devices: anomalously high (almost infinite) subthreshold slope and reduction of the drain breakdown voltage.

5.7.1. Anomalous subthreshold slope

As we have seen at the end of Section 5.6.1, the generation of majority carriers (holes in the case of an n-channel transistor) by impact ionization near the drain can give rise to an increase of the body potential and decrease of the threshold voltage. Sometimes, a similar effect can occur at gate voltages lower than the threshold voltage . If the drain voltage is high enough, impact ionization can occur in the subthreshold region, even though the drain current is very small. This effect is observed in partially depleted devices and in fully depleted devices with back accumulation, and can be explained as follows. When the device is turned off, there is no impact ionization, and the body potential is equal to zero, since there is no base-to-source current. When the gate voltage is increased the weak inversion current can induce impact ionization in the high electric field region near the drain, holes are generated, the body potential increases, and the threshold voltage is reduced. Consequently, the whole $I_D(V_{GS})$ characteristic shifts to the left, and the current can increase with gate voltage with a slope larger than 60 mV/decade. In other words, inverse subthreshold slopes lower than the theoretical limit (in absence of multiplication effect) of 60 mV/decade, can be observed [5.41, 5.53].

If the minority carrier lifetime in the silicon film is high enough, the parasitic bipolar transistor present in the NPN structure of the MOS device can amplify the base current (*i.e.* amplify the hole current generated by impact ionization near the drain). The base current is given by I_{body} = (M-1) $I_{Dsat} \equiv$ (M-1) I_{ch} (see Section 5.6) where I_{ch} is the channel current, and the resulting increase of drain current is given by $\Delta I_D = \beta\ I_{body} = \beta$ (M-1) I_{ch} (Figure 5.7.1). This increase of drain current constitutes a positive feedback loop on the current flowing through the device: the drain current suddenly increases, and an infinite subthreshold slope is observed (Figure 5.7.2).

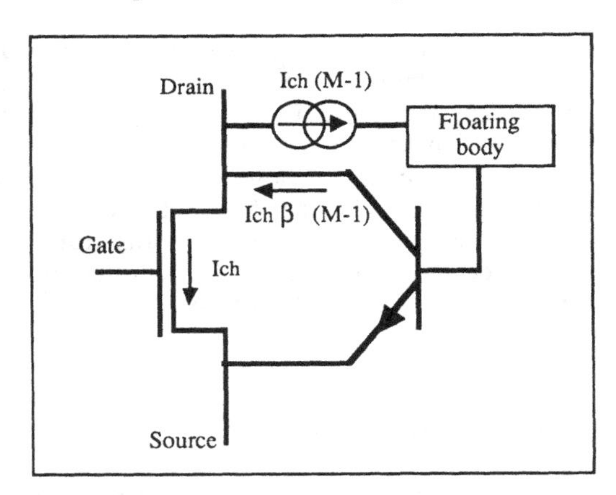

Figure 5.7.1: Parasitic bipolar transistor of the SOI MOSFET. I_{ch} is the channel current.

This phenomenon is known as the "single-transistor latchup" [5.42-5.45]. It can be explained as follows (Figure 5.7.2). At low drain bias (curve a), a normal subthreshold slope is obtained, under forward as well as ar reverse gate voltage scan conditions. If the drain voltage is increased (curve b), the impact ionization near the drain raises the body potential (forward gate voltage scan case). This reduces the threshold voltage and leads to an increase of the drain current, which in turn increases impact ionization near the drain. When the gain of the positive feedback loop, β (M-1), reaches unity, the current suddenly increases. The positive feedback is self-limiting: increased body bias also increases the drain saturation voltage which results in a lower electric field near the drain and a smaller impact ionization current. During a descending gate voltage scan (curve b), the impact ionization current under a large drain voltage keeps the body voltage high, which in turn keeps the threshold voltage of the device low, and a high drain current is observed until the positive feedback can no longer be maintained, so that β (M-1) < 1, and the drain current suddenly drops. It can be noticed that the gate voltage at which the current suddenly raises during a forward scan is higher of sudden increase the voltage at which the current suddenly falls during a reverse scan. As a consequence, an hysteresis is observed in the $I_D(V_G)$ curve (curve b). If the drain bias is large enough, the positive feedback loop cannot be turned off once it has been triggered (β (M-1) remains larger or equal to 1), and the device cannot be turned off (curve c). It is worth noting that this parasitic phenomenon was not observed in devices made in early SOI material, where the recombination of minority carriers in the base was low and where the gain of the bipolar, β, was small. The single-transistor latch effect is not observed if the body of the device is grounded, and it is reduced if the device is fully depleted.

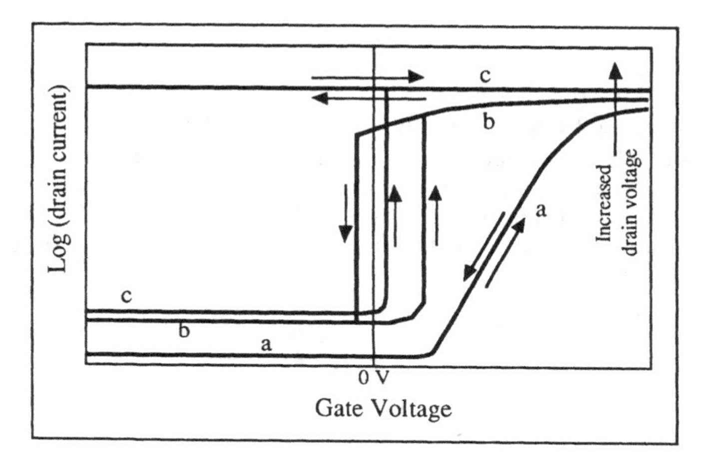

Figure 5.7.2: Illustration of the single-transistor latch [5.43]. "Normal" subthreshold slope at low drain voltage (a), infinite subthreshold slope and hysteresis (b), and device "latch-up" (c).

5.7.2. Reduced drain breakdown voltage

It can been demonstrated that the peak electrical field at the drain junction of SOI devices is lower than what is commonly found in bulk devices if the junction is reaching through the silicon film to the buried oxide [5.46]. From this result, a higher junction breakdown voltage could be expected in SOI transistors than in bulk devices. Unfortunately, the SOI MOSFET presents a parasitic bipolar transistor with floating base. From bipolar transistor theory [5.47], the collector (drain) breakdown voltage with open base, BV_{CEO}, is smaller than when the base is grounded (BV_{CBO}). Both breakdown voltages relate as follows:

$$BV_{CEO} = \frac{BV_{CBO}}{\sqrt[n]{\beta}} \qquad (5.7.1)$$

where β is the gain of the bipolar device, and *n* ranges typically between 3 and 6 [5.47]. The above relationship is quite a simplification of the breakdown mechanisms occurring in SOI MOSFETs, since both β and M-1 (the multiplication factor) depend on the drain voltage in a highly nonlinear fashion [5.42]. From bipolar transistor theory [5.47], one can also write:

$$\beta \cong 2\,(L_n/L_B)^2 - 1 \qquad (5.7.2)$$

where L_B is the base width, which can be assumed, in first approximation, to be equal to the effective channel length, L, and L_n is the electron diffusion length: $L_n{}^2 = D_n\,\tau_n$, where D_n and τ_n are the diffusion coefficient and the lifetime of the minority carriers in the base, respectively. From (5.7.1) and (5.7.2), and taking BV_{CEO} as the drain breakdown voltage with floating body, equal to BV_{DS}, one obtains:

$$BV_{DS} = \frac{BV_{CBO}}{\sqrt[n]{\frac{2 D_n \tau_n}{L^2} - 1}} \qquad (5.7.3)$$

where L is the gate length.

Using Einstein's relationship $D_n = \frac{kT}{q}\mu_n$, an abacus relating BV_{DS}/BV_{CBO} to channel length can be plotted, with the minority carrier lifetime in the SOI material as parameter (Figure 5.7.3) [5.48]. It should however be noted that equations (5.7.2) and (5.7.3), and, hence, Figure 5.7.3 are valid only if $L_n > L$, *i.e.* if β is significantly larger than 1, (*i.e.* when short channels and relatively good lifetime SOI material are used.)

This reduction of the drain breakdown voltage was not observed in devices made in early (and defective) SOI material, where the minority carrier lifetime was low, and its control constitutes one of the major challenge for today's SOI research activity [5.49], especially when submicron devices are considered. Possible solutions of the problem include the use of lightly-doped sources and drains (LDS and LDD), use of lifetime killers, controlled introduction of defects in the silicon film, and the use of body contacts [5.48]. Finally, it should be noted that the above model stems on the presence of a neutral base region, and is, therefore, valid only for partially depleted devices. The understanding of parasitic bipolar effects in fully depleted devices necessitates the development of a model of the bipolar transistor with fully depleted base [5.13, 5.42]. Numerical simulations show that, even though no kink is discernible in fully depleted SOI MOSFETs, these devices are subjected to significant body charging by impact ionization, which portends significant parasitic bipolar effects [5.42]. The drain breakdown voltage is controlled by the common-emitter bipolar breakdown voltage BV_{CEO} which occurs under the condition that $\beta(M-1)=1$. Both β and M are highly nonlinear. M increases strongly with V_{DS} due to the increasing drain electric field, and β decreases strongly with V_{DS} due to the prevalent high-injection in the body (base of the bipolar device) [5.42]. Numerical analysis also suggests that the use of an LDS structure can reduce β through the reduction of the bipolar emitter efficiency and concludes that an optimized LDD/LDS fully depleted SOI MOSFET renders submicron SOI CMOS a viable and advantageous technology [5.42].

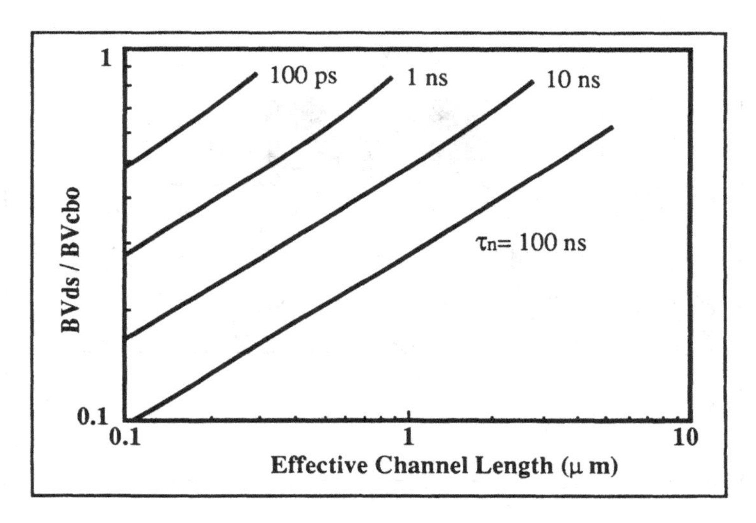

Figure 5.7.3: Relationship between the reduction of drain breakdown voltage, gate length and minority carrier lifetime (τ_n) in the SOI material.

5.8. Accumulation-mode p-channel MOSFET

Thick-film p-channel MOSFETs (with an N^+ poly gate) are usually buried-channel devices, and their characteristics are similar to those of bulk buried-channel pMOSFETs.

Thin-film, enhancement-mode p-channel MOSFETs can also fabricated [5.16] but these devices have exhibit large values of threshold voltage when they have a thin (20 nm or less) gate oxide and N^+ polysilicon as gate material. In order to to obtain useful values of threshold voltage (around -0.7 V) when a thin gate oxide and an N^+ polysilicon gate are used, the body of the transistor has to have a p-type doping (Figure 5.8.1). Such a device is an accumulation-mode device (also called deep-depletion device), the characteristics of which will be derived in this section.

When the device is turned OFF, the silicon film is fully depleted due to the presence of positive interface charges and to the negative value of the work function difference between the N+ polysilicon gate to the P-type body of the device. When the device is turned ON, the film is no longer fully depleted and conduction occurs both in the body of the device and in a surface accumulation channel [5.50].

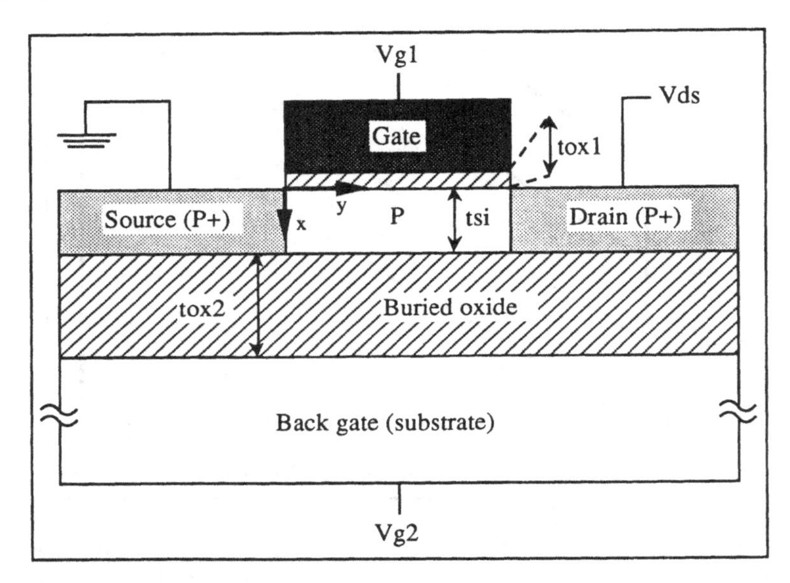

Figure 5.8.1: Cross-section of a p-channel SOI MOSFET illustrating some of the notations used in this Section.

When a zero bias is applied to the gate, the film is fully depleted due to both the presence of positive charges at the $Si\text{-}SiO_2$ interface and to that of the use of an N+-polysilicon gate. The gate material-silicon work function difference is given by:

$$\Phi_{MS1} = -\frac{E_g}{2} - \frac{kT}{q}\ln\frac{N_a}{n_i} \tag{5.8.1}$$

where n_i is the intrinsic carrier concentration in silicon. When a negative gate voltage is applied, the overall hole concentration in the silicon film is increased. Threshold of accumulation is reached when $\Phi_{s1}=0$. The gate voltage needed to accomplish this is given by:

$$V_{th,acc} = \Phi_{MS1} - \frac{Q_{ox1}}{C_{ox1}} = V_{fb1} \tag{5.8.2}$$

Where V_{fb1} is the front flat-band voltage, Q_{ox1} is the areal charge density in the gate oxide, and C_{ox1} is the gate oxide capacitance. When the gate voltage, V_{G1}, is lower (i.e. larger in absolute value) than the accumulation threshold voltage, the accumulation charge in the channel is given by: $Q_{acc}(y) = -[V_{G1} - V_{fb1} - V(y)]\, C_{ox1}$, where V(y) is the local potential along the channel (y=0 at the source junction and y=L at the drain junction). Using the gradual channel approximation, one obtains the following expression for the current in the accumulation channel:

$$\int_0^{V_{DS}} dV = I_{acc} \int_0^{L} dR(y) \quad \text{, with } dR(y) = \frac{dy}{W\, \mu_s\, Q_{acc}} \tag{5.8.3}$$

which yields after integration:

$$I_{acc} = \frac{W}{L} \mu_s C_{ox1} \left((V_{G1} - V_{fb1})\, V_{DS} - \frac{V_{DS}^2}{2} \right) \tag{5.8.4}$$

in the linear regime ($V_{DS} > V_{G1} - V_{fb1}$), and

$$I_{acc} = \frac{W}{L}\, \mu_s \frac{C_{ox1}}{2} (V_{G1} - V_{fb1})^2 \tag{5.8.5}$$

above saturation ($V_{DS} < V_{G1} - V_{fb1}$).

μ_s is the hole surface mobility, which is equal to $\frac{\mu_{s0}}{1+\theta(V_{fb1}-V_{G1})}$, with μ_{s0} being the zero-field surface mobility, and θ being a field mobility reduction factor.

The other current flowing from source to drain is the body current, which appears if there is a portion of non-depleted silicon below the channel. As far as this current component is concerned, similarity is found between the operation of this MOS device and that of a JFET. Conduction between source and drain occurs in the neutral (undepleted) part of the body. The width of this conduction path is modulated by the vertical extension of the depletion zones related to the front and the back gate. The front gate depletion depth can be found by solving the Poisson equation using the depletion approximation:

$$\frac{d^2\Phi(x)}{dx^2} = \frac{q\,N_a}{\varepsilon_{si}} \qquad (5.8.6)$$

with $\Phi(x)$ being the potential in the x-direction (in the depth of the semiconductor film), and N_a the acceptor doping concentration.
We find:

$$-E(x)= \frac{q\,N_a\,x}{\varepsilon_{si}} + A \quad \text{and} \quad \Phi(x) = \frac{q\,N_a\,x^2}{2\varepsilon_{si}} + Ax + B$$

(A and B are integration constants) with the following boundary conditions: $E(x_{depl})=0$ and $\Phi(x_{depl})=0$ where x_{depl} is the depletion depth. From the above expressions, we find the surface potential:

$$\Phi(0) = \frac{q\,N_a\,x_{depl}^2}{2\varepsilon_{si}}$$

Using Gauss' law, we finally obtain:

$$V_G - V_{FB} = \frac{q\,N_a\,x_{depl}^2}{2\varepsilon_{si}} + \frac{q\,N_a\,x_{depl}}{C_{ox}} \qquad (5.8.7)$$

The gate voltage for which the depletion depth equals the silicon thickness is:

$$V_G = V_{FB} + \frac{q\,N_a\,t_{si}^2}{2\varepsilon_{si}} + \frac{q\,N_a\,t_{si}}{C_{ox}} \equiv V_{FB} + V_{depl} \qquad (5.8.8)$$

V_{depl} being defined by the above relationship (5.8.8).

As long as there is a portion of neutral silicon in the film (at a distance *y* from the source), the depth of the front-gate-related depletion region is given by:

$$X_{d1}(y) = \frac{-\varepsilon_{si}}{C_{ox1}} + \sqrt{\varepsilon_{si}^2/C_{ox1}^2 + 2\,\varepsilon_{si}\,(V_{G1} - V_{fb1} - V(y))/q\,N_a} \qquad (5.8.9)$$

Similarly, under the same conditions, the depth of the depletion zone arising from the back interface is given by:

$$X_{d2}(y) = \frac{-\varepsilon_{si}}{C_{ox2}} + \sqrt{\varepsilon_{si}^2/C_{ox2}^2 + 2\,\varepsilon_{si}\,(V_{G2} - V_{fb2} - V(y))/q\,N_a} \qquad (5.8.10)$$

If $X_{d1}(y) + X_{d2}(y)$ is equal to or larger than t_{si}, the silicon film thickness, the film is locally fully depleted. For the sake of simplicity, it will be assumed that the width of the back depletion region, X_{d2}, is constant and given by equation (5.8.10) where V(y) is held constant and equal to V(y=0). The resistance of an elementary resistor in the body channel is given by:

$$dR = \frac{dy}{W\,\mu_b\,q\,N_a\,(t_{eff} - X_{d1})} \qquad (5.8.11)$$

where t_{eff} is equal to t_{si}-X_{d2} and μ_b is the bulk hole mobility. Using again the gradual channel approximation, and integrating dV=IdR from source to drain for the various operation modes of the device, and using Eqn. (5.8.11) to find R and Eqn. (5.8.9) to calculate X_{d1}, one obtains the following expression:

$$I_{body} \int dy = W\, q\, Na\, \mu_b \int [t_{eff} + \frac{\varepsilon_{si}}{Cox1} - \sqrt{\frac{\varepsilon_{si}^2}{Cox1^2} + \frac{2\varepsilon_{si}\,(VG1 - Vfb1 - V(y))}{q\, Na}}\;]\; dV \quad (5.8.12)$$

The integration of Equation (5.8.12) from source to drain gives an expression in the form:

$$I_{body} = q\, N_a\, \mu_b \frac{W}{L} \alpha$$

where α results from the integration of (5.8.12) and varies as a function of applied biases, depending whether the film is fully depleted or not, whether the neutral zone extends all the way to the drain, etc... We will also define V'_{depl} as the V_{depl} of equation (5.8.8) where t_{si}=t_{eff}, or:

$$V'_{depl} \equiv \frac{q\, N_a\, t_{eff}^2}{2\varepsilon_{si}} + \frac{q\, N_a\, t_{eff}}{C_{ox}}$$

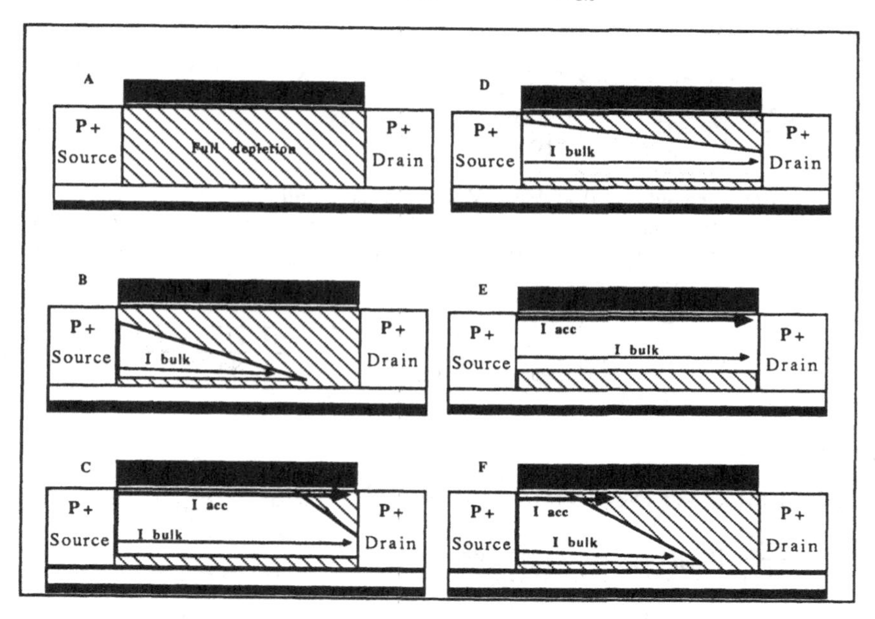

Figure 5.8.2: Cross-section of a p-channel SOI MOSFET illustrating the possible distributions of depletion and neutral zones in the silicon film.

The following cases can be distinguished:

* $\alpha = 0$ when $V_{G1} > V_{fb1} + V'_{depl}$, *i.e.* when the film if fully depleted. V'_{depl} is the change in front gate voltage with respect to flatband necessary to obtain $X_{d1} = t_{eff}$ near the source (V(y=0)) using equation (5.8.9). There is no accumulation channel (Figure 5.8.2.A).

* $\alpha = t_{eff} V_{DS}$ when $V_{G1} - V_{fb1} < 0$ and $V_{G1} - V_{fb1} - V_{DS} < 0$ (Figure 5.8.2.E), *i.e.* when the front interface is in accumulation from source to drain.

* In the case where $0 < V_{G1}-V_{fb1} < V'_{depl}$ and $0 < V_{G1}-V_{fb1}-V_{DS} < V'_{depl}$, *i.e.* when neither accumulation nor body channel pinchoff occur (Figure 5.8.2.D), equation (5.8.12) can be written:

$$I_{body} \int_0^L dy = W\, q\, N_a\, \mu_b \int_0^{V_{DS}} [t_{eff} + \frac{\varepsilon_{si}}{C_{ox1}} - \sqrt{\frac{\varepsilon_{si}^2}{C_{ox1}^2} + \frac{2\varepsilon_{si}\,(VG1-Vfb1-V(y))}{q\,Na}}\,]\; dV$$

and, therefore:

$$\alpha = (t_{eff} + \varepsilon_{si}/C_{ox1})\, V_{DS} + \frac{qN_a}{3\varepsilon_{si}} [(\varepsilon_{si}/C_{ox1})^2 + \frac{2\varepsilon_{si}}{qN_a} (V_{G1}-V_{fb1}-V_{DS})]^{3/2} - \frac{qN_a}{3\varepsilon_{si}} [(\varepsilon_{si}/C_{ox1})^2 + \frac{2\varepsilon_{si}}{qN_a} (V_{G1}-V_{fb1})]^{3/2} \qquad (5.8.13)$$

* In the case where $0 < V_{G1}-V_{fb1} < V'_{depl}$ and $V_{G1}-V_{fb1}-V_{DS} > V'_{depl}$, i.e. when body channel is pinched off, but there is no accumulation at the source end nor full depletion of the film (Figure 5.8.2.B), we have:

$$I_{body} \int_0^L dy = WqNa\, \mu_b \int_0^{V_{G1}-V_{FB1}-V'_{depl}} [t_{eff} + \frac{\varepsilon_{si}}{C_{ox1}} - \sqrt{\frac{\varepsilon_{si}^2}{C_{ox1}^2} + \frac{2\varepsilon_{si}\,(VG1-Vfb1-V(y))}{q\,Na}}\,]\; dV$$

which yields:

$$\alpha = (t_{eff} + \varepsilon_{si}/C_{ox1})\, (V_{G1}-V_{fb1}-V'_{depl}) + \frac{qN_a}{3\varepsilon_{si}} [(\varepsilon_{si}/C_{ox1})^2 + \frac{2\varepsilon_{si}}{qN_a} V'_{depl}]^{3/2} - \frac{qN_a}{3\varepsilon_{si}} [(\varepsilon_{si}/C_{ox1})^2 + \frac{2\varepsilon_{si}}{qN_a} (V_{G1}-V_{fb1})]^{3/2} \qquad (5.8.14)$$

* In the case where $0 > V_{G1}-V_{fb1}$ and $0 < V_{G1}-V_{fb1}-V_{DS} < V'_{depl}$, *i.e.* when body channel is not pinched off and there is accumulation at the source end (Figure 5.8.2.C), we have:

$$I_{body}\int_0^L dy = WqNa\,\mu_b \int_{V_{G1}-V_{FB1}}^{V_{DS}} \left[t_{eff} + \frac{\varepsilon_{si}}{C_{ox1}} - \sqrt{\frac{\varepsilon_{si}^2}{C_{ox1}^2} + \frac{2\varepsilon_{si}\,(VG1-Vfb1-V(y))}{q\,Na}}\,\right] dV$$

$$+\; WqNa\,\mu_b \int_0^{V_{G1}-V_{FB1}} t_{eff}\, dV$$

which yields:

$$\alpha = t_{eff}\,(V_{G1}-V_{fb1}) + (t_{eff} + \varepsilon_{si}/C_{ox1})\,(V_{DS}-V_{G1}+V_{fb1}) + \frac{qN_a}{3\varepsilon_{si}}\left[(\varepsilon_{si}/C_{ox1})^2 + \frac{2\varepsilon_{si}}{qN_a}(V_{G1}-V_{fb1}-V_{DS})\right]^{3/2} - \frac{qN_a}{3\varepsilon_{si}}\,[\varepsilon_{si}/C_{ox1}]^3 \qquad (5.8.15)$$

* In the case where $0 > V_{G1}-V_{fb1}$ and $V_{G1}-V_{fb1}-V_{DS} > V'_{depl}$, *i.e.* when body channel is pinched off and there is accumulation at the source end. (Figure 5.8.2.F), we have:

$$I_{body}\int_0^L dy = WqNa\mu_b \int_{V_{G1}-V_{FB1}}^{V_{G1}-V_{FB1}-V'_{depl}} \left[t_{eff} + \frac{\varepsilon_{si}}{C_{ox1}} - \sqrt{\frac{\varepsilon_{si}^2}{C_{ox1}^2} + \frac{2\varepsilon_{si}\,(VG1-Vfb1-V(y))}{q\,Na}}\,\right] dV$$

$$+\; WqNa\,\mu_b \int_0^{V_{G1}-V_{FB1}} t_{eff}\, dV$$

which yields:

$$\alpha = t_{eff}\,(V_{G1}-V_{fb1}) - (t_{eff} + \varepsilon_{si}/C_{ox1})\,(V'_{depl}) + \frac{qN_a}{3\varepsilon_{si}}\left[(\varepsilon_{si}/C_{ox1})^2 + \frac{2\varepsilon_{si}}{qN_a}V'_{depl}\right]^{3/2} - \frac{qN_a}{3\varepsilon_{si}}\,[\varepsilon_{si}/C_{ox1}]^3 \qquad (5.8.16)$$

Finally, the total drain current in the device is given by the sum of the current in the accumulation channel and the current in the body of the device:

$$I_{DS} = I_{acc} + I_{body} \qquad (5.8.17)$$

Figure 5.8.3 presents the linear $I_D(V_{G1})$ characteristics of a thin-film, accumulation-mode p-channel transistor for different back-gate voltages. The *apparent* front threshold voltage of this device shows a dependence on back-gate bias which is similar to that which is observed in fully-depleted enhancement-mode devices. Actually, the front accumulation threshold voltage is quite independent of back-gate bias, and the shift of the $I_D(V_{G1})$ characteristics with (more negative) back-gate bias is due to the apparition and increase of the body current. A large negative back biases, an accumulation channel can even be created at the back interface. This case, however, was not dealt with in the above model, since the presence of a back accumulation channel is undesirable for most practical applications.

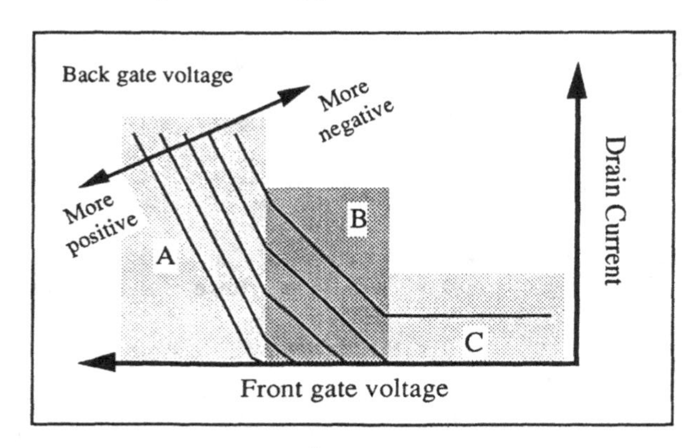

Figure 5.8.3: Linear $I_D(V_{G1})$ characteristics of a thin-film, accumulation-mode p-channel transistor for different back-gate voltages. The front accumulation current, the body current, and the back channel accumulation current are outlined by the shaded zones A, B, and C, respectively.

Figure 5.8.4 presents the output characteristics of a thin-film, accumulation-mode p-channel SOI device (t_{si}=100 nm, N_a=4 10^{16} cm^{-3}).

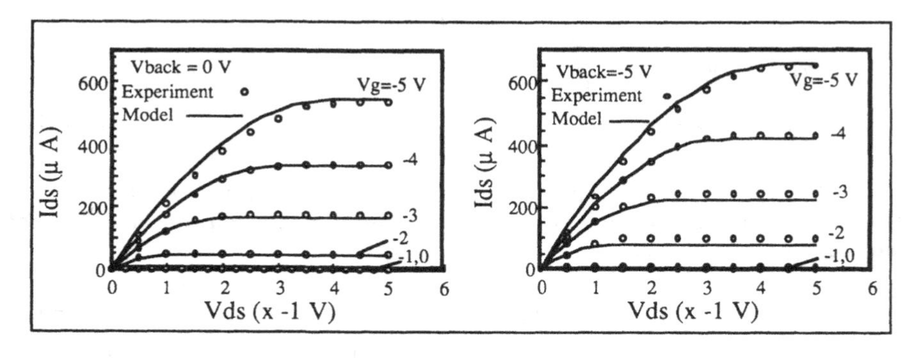

Figure 5.8.4: Output characteristics of a thin-film, accumulation-mode p-channel SOI device, for two value of back-gate bias.

It is worth noting that in most applications, p-channel SOI MOSFETs are operating with a negative back-gate bias. If one considers, for instance, the case of a CMOS inverter, the source of the p-channel device is connected to $+V_{DD}$ (*e.g.* 5 volts). If the mechanical substrate (the back gate) is grounded, the p-channel transistor is then operating with a back-gate bias equal to $-V_{DD}$ (*i.e.* -5 V).

Good performances are usually obtained from accumulation-mode p-channel MOSFETs. Because of the reduced scattering encountered by the bulk current, a higher mobility is observed in these devices than in enhancement-mode p-channel MOSFETs [5.11].

Accumulation-mode ($N^+N^-N^+$) n-channel MOSFETs can also be fabricated, using either p^+ polysilicon or n^+ polysilicon as gate material [5.2, 5.51]. In the latter case, however, devices with negative threshold voltage are obtained. Either devices can be simulated using the above model, provided the proper sign changes are made.

CHAPTER 6 - Other SOI Devices

Although CMOS remains the most obvious field of application for SOI, the ease of processing SOI substrates, the full dielectric isolation of the devices and the possibility of using a back gate have sparked a large research activity in the field of novel SOI devices. Indeed, different novel bipolar and MOS structures have been proposed such as gated diode structures, lateral bipolar and bipolar-MOS devices, vertical bipolar transistors with back gate-induced collector, high-voltage lateral devices of various kinds and double gate MOS devices. This Chapter will review these devices, qualitatively explain their physics and explore their possible fields of application. It will also describe some devices (other than the MOSFET) which have been adapted from bulk to SOI technology.

6.1. Non-conventional devices adapted from bulk

Thick SOI films, such as those produced by the bonding and etch back technique, can be used as a replacement for the DI process [6.1] or to fabricate complementary bipolar circuits [6.2]. Smart power SOI ICs, combining low-voltage CMOS logic and power devices have been realized using a 3D approach [6.3]. Several CMOS/SOI-compatible lateral high-voltage SOI devices have been reported [6.4-6.6]. All these devices result from adaptation of existing bulk silicon structures to SOI technology, and only three examples of such adapted devices will be described here: the lateral SOI COMFET, the high-voltage lateral MOSFET, the lateral PIN diode, and the JFET.

6.1.1. COMFET

SOI lateral COMFETs (can be realized in relatively thin SOI material using a standard SOI CMOS process (Figure 6.1.1)[6.4]. Despite the use of such a thin gate oxide as 23 nm, these devices have forward and reverse breakdown voltages of 80 volts. When the device is turned off (low gate voltage), the reverse-biased diode between the P-type region under the gate and the N^- drift region prevent current from flowing through the structure. The low dopant concentration in the N^- region ensures a relatively high breakdown voltage. When the gate is positively biased, electrons are injected from the N^+ part of the cathode into the N^- region through a channel created underneath the gate oxide in the p-type region. These electrons can recombine with holes injected in the N^- region from the positively biased anode. The devices then basically operates as a forward-biased PIN diode, and current flows through the structure. The limitation of these thin-film devices (a silicon film thickness of 200 nm was used in the reported case) lies in the fact that rather high current densities are readily reached in the silicon film, limiting the usefulness of such devices for power applications. Their ability to control high voltages and their compatibility with CMOS processing, on the other hand, may render them useful for applications where high-speed SOI CMOS logic and relatively high-voltage interface devices are required.

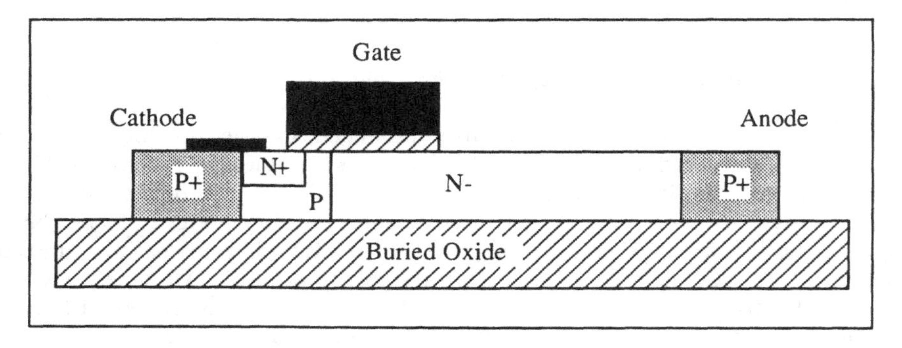

Figure 6.1.1: Lateral SOI COMFET.

6.1.2. High-voltage lateral MOSFET

Double implantation of oxygen ions has recently been employed to produce CMOS and high-voltage, offset-gate MOS devices (Figure 6.1.2). The double implantation is performed as follows: after the formation of classical SIMOX material, epitaxy is used to increase the thickness of the silicon overlayer. The epitaxial layer then serves as substrate for the formation of a second SIMOX structure, in such a way that a "silicon-on-oxide-on-silicon-on-oxide" structure is obtained. The buried silicon layer can be connected to ground and serves as a buried back gate which acts as a shield from electrical interference from high-voltage devices realized in the substrate. Masked implantation can indeed be used to create devices in both the bulk silicon (which is locally protected from oxygen implantation by a mask) and in the SOI overlayer [6.6]. A second role of the buried gate is to reduce the maximum electric field in the device (RESURF device) by providing a grounded field plate under the transistor. As in most high-voltage lateral MOSFETs, the function of the N-type drift region near the drain is to reduce the drain electric field between and, hence, to improve the breakdown voltage. Breakdown voltages of 90 volts are indeed obtained for an active silicon thickness of 400 nm and a gate oxide thickness of 100 nm.

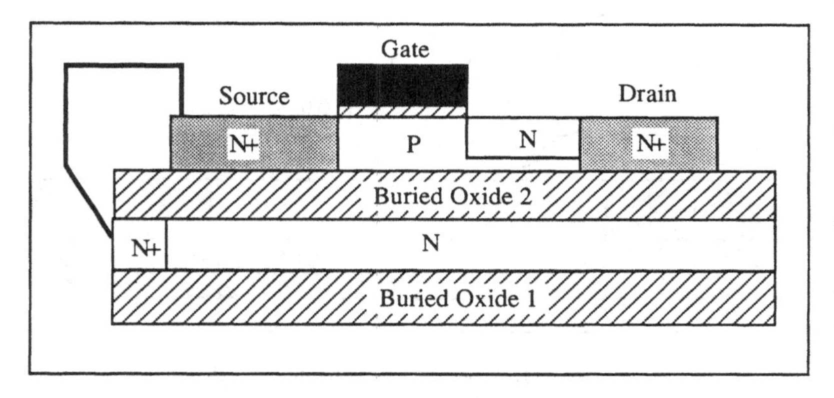

Figure 6.1.2: High-voltage SOI MOSFET [6.6].

6.1.3. PIN photodiode

PIN diodes can easily be produced using an SOI CMOS process without addition of any mask step. The intrinsic silicon can either be protected from threshold voltage implants or be lightly doped (by threshold voltage adjust implants). The N+ cathode and the P+ anode are formed using the n-channel and p-channel device source and drain implants. PIN diodes with reverse breakdown voltages larger than 50 V can readily be formed, even in thin SOI films. PIN diodes can be useful as ESD input protection devices. Reverse-biased thin-film PIN diodes can also be used as photosensors (Figure 6.1.3)[6.7]. Indeed, electron-hole pairs are created under illumination within the large, fully depleted intrinsic region. The carriers are separated by the electric field in that region and collected by the N^+ and P^+ diffusions. Because the devices are made in a thin silicon film which is part of a multilayer structure (oxide passivation, silicon film, buried oxide and silicon substrate), light interferences occur and, as a result, the structure absorbs more some wavelengths than others. In general red light is not absorbed (the silicon film is too thin), but peaks of absorption can be obtained in the blue and green regions of the visible spectrum.

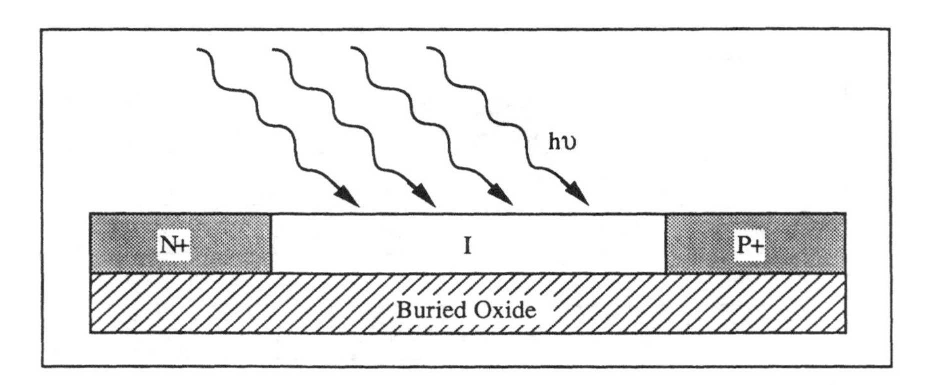

Figure 6.1.3: Lateral SOI PIN photodiode.

6.1.4. JFET

Junction field-effect transistors (JFETs) can also be fabricated on SOI substrates. JFETs are particularly interesting for applications where low noise, high input impedance and good radiation hardness are required. Figure 6.1.4 shows the cross-section of an SOI p-channel JFET [6.24]. The structure is fabricated as follows. Buried N^+ diffusions are created in the silicon overlayer of a SIMOX wafer, and p-type silicon epitaxy is carried out to increase the silicon film thickness from 0.2 to 1 μm. These buried N^+ diffusions will be utilized as back gate. The silicon islands are defined by means of a mesa isolation process, and ion implantation is used to form the top front-gate (N^+) and source and drain (P^+) electrodes (Figure 6.1.4). Modulation of the source-drain current can be achieved by applying a voltage to either the front gate or the back gate while keeping the other gate grounded, but the highest transconductance is obtained by connecting G1 and G2 together and using them as a single input terminal.

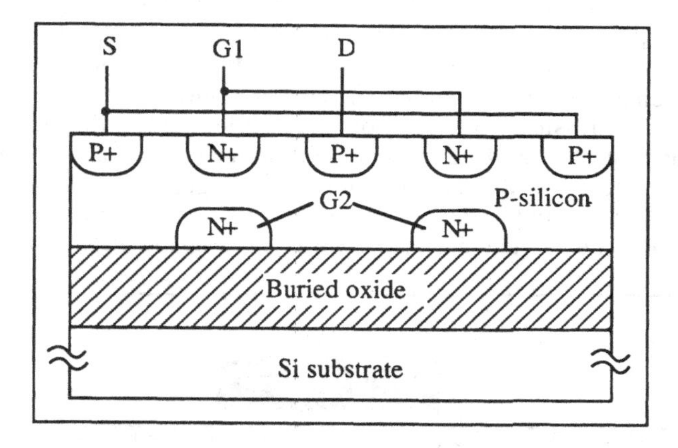

Figure 6.1.4: P-channel JFET. G1 is the front gate, G2 is the back gate.

SOI JFETs show good radiation hardness characteristics. Indeed, upon dose irradiation (X-rays), a saturation current degradation of 20% is observed for a 30 krad(Si) irradiation. Above 30 krad(Si) and up to 1 Mrad(Si), no further shift is observed. The current saturation degradation caused by a 1 MeV neutron fluence of 10^{14} cm^{-2} is below 10% [6.24].

6.2. Novel SOI devices

The flexibility of SOI technology extends, however, beyond the adaptation of existing devices to SOI substrates. Novel device possibilities have recently been demonstrated, which can hardly or not be made in bulk silicon. SOI technology indeed offers two unique possibilities: firstly, full dielectric isolation of the devices is achieved, which permits one to fabricate novel devices without the burden of parasitic effects such as latch-up and without having undesired forward-biased junctions or leakage current paths between devices. Secondly, the SOI structure offers the possibility of using a second gate at the back of the device. We will now review some of these devices in more detail.

6.2.1. Lubistor

The lateral unidirectional bipolar-type insulated-gate transistor or Lubistor (Figure 6.2.1) is a gated $N^+N^-P^+$ (or $N^+P^-P^+$) diode which exhibits triode characteristics and is able to carry current densities up to 10^5 A/cm^2. It has first been realized in SIMOX material [6.8], although the intrinsic properties of the device are quite independent of the SOI material being used. The operation principle of the device is the following: a positive anode to cathode voltage, V_{AK} is applied, and a positive gate to cathode voltage, V_{GK}, is used to control the current flow in the device. The device basically works as a gated PIN diode.

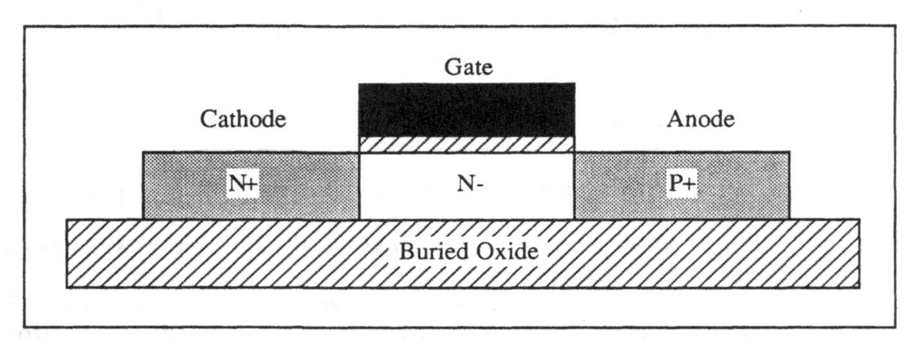

Figure 6.2.1: Lateral unidirectional bipolar-type insulated-gate transistor (Lubistor).

The Lubistor can be fully turned off by the gate if the silicon film thickness, t_{si}, is smaller than the Debye length, L_{DE}, defined by: $L_{DE}=\sqrt{\frac{2\varepsilon_{si}kT}{qN_d}}$, where N_d is the dopant concentration in the N^- region. In the ON state, the device displays output characteristics in the form: $I_{AK}=A[V_{AK}-B(V_{GK}-V_{FB})]^n$, where A and B are parametric constants, V_{FB} is the flat-band voltage in the N^- region, and $2 < n < 3$ (Figure 6.2.2).

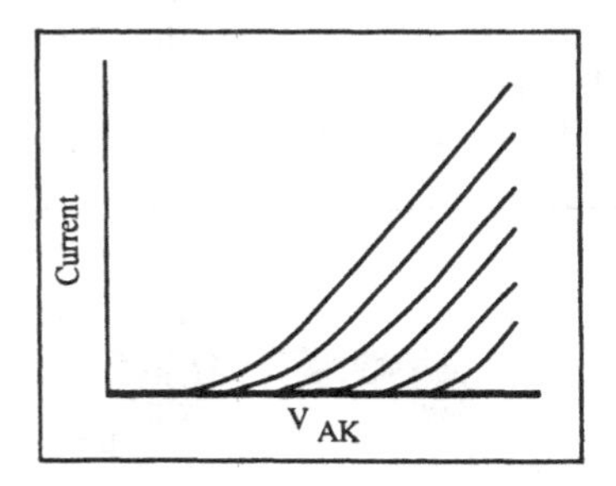

Figure 6.2.2: Triode-like output characteristics of the Lubistor.

In the OFF state, the potential of the thin N^- region is higher than that of both the cathode and the anode, and carrier injection does not take place in spite of the fact than the anode bias is higher than the built-in potential of the anode P-N junction (Figure 6.2.3). In the ON state, the anode potential is higher than that of the N^- region, and holes are injected into the N^- region. Electrons are also injected in the same region from the cathode, and the device behaves like a forward-biased PIN diode (Figure 6.2.3). Due to its high current drive capability, the Lubistor has potential as an output buffer device.

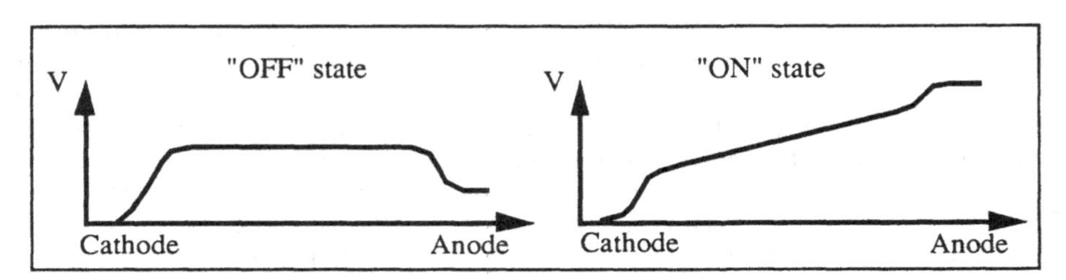

Figure 6.2.3: Anode-to-cathode potential distribution in the Lubistor in the ON and the OFF state.

6.2.2. Bipolar-MOS device

As seen in Section 5.7, every enhancement-mode SOI MOSFET contains a parasitic bipolar transistor. The voltage-controlled bipolar-MOS (VCBM) device controls the bipolar effect and makes use of the combined current drive capabilities of both the bipolar and the MOS parts of a "normally MOS" device [6.9]. This is achieved by connecting the gate, which controls the current flow in the MOS part of the device, to the floating substrate, which acts as the base of the lateral bipolar transistor (Figure 6.2.4). The source and the drain of the MOS transistor are also the emitter and the collector of the bipolar device, respectively.

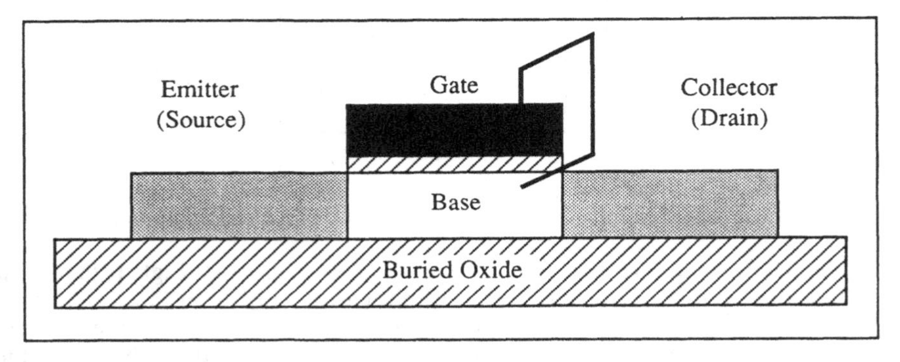

Figure 6.2.4: Voltage-controlled bipolar-MOS device (VCBM).

We will now consider the case of an n-channel (NPN) device, although p-channel (PNP) devices can be realized as well). When the device is OFF ($V_G=V_B=0$), the potential of the MOS substrate (= the base) is low, which maximizes the value of the threshold voltage (and thereby minimizes the OFF current). When a gate bias is applied, on the other hand, the potential of the MOS substrate is increased, which decreases the threshold voltage according to the expressions developed in section 5.3.2 for the body effect (bulk case). This lowering of the threshold voltage increases the current drive for a given gate voltage, compared to an MOS transistor without gate-to-body connection. Similarly, the application of a gate voltage when the device is "ON" increases the collection efficiency of the bipolar device, and reduces the effective neutral base width of the device, thereby improving the bipolar gain. This effect is illustrated in Figure 6.2.5. In absence of any gate bias (we assume a flat-band condition at the Si-SiO_2 interfaces), the minority carrier (electrons) current injected from the emitter has to diffuse across the width of the neutral base before being collected by the depletion zone near the collector (Figure 6.2.5.A). According to equation 5.7.2, the gain of the bipolar transistor, β, can be approximated by: $\beta \cong 2(L_n/L_B)^2 - 1$, where L_n is the minority carrier (electrons in this case) diffusion length in the base, and L_B is the base width. When a positive gate voltage is applied, an inversion channel is created. This channel allows for the MOS current component , I_{ch}, to flow from source to drain. Two components of bipolar current can be distinguished: the electrons injected by the emitter can either be attracted by the surface potential created by the gate bias and join the channel electrons (I_{C1} component) or diffuse directly from emitter to collector through the base (I_{C2} component). In both cases, the path traveled by the electrons is shorter than if no gate bias were applied (Figure 6.2.5.B). Indeed, the widening of the depletion zone caused by the gate bias reduces the effective neutral base length. The even shorter path traveled by the electrons injected from the emitter and joining the channel current is also obvious (Figure 6.2.5.B). As a result from these considerations and of equation 5.7.2, the gain of the bipolar device is increased when a positive gate voltage is applied, which is always the case in the VCBM device, since the gate is connected to the base. Recently, this principle of gain enhancement has been applied to bulk lateral bipolar transistors [6.10]. In summary, the presence of a gate improves the gain of the bipolar transistor, and the presence of a base contact improves both the ON and OFF characteristics of the device. Such a mutually beneficial phenomenon can almost be called "symbiosis".

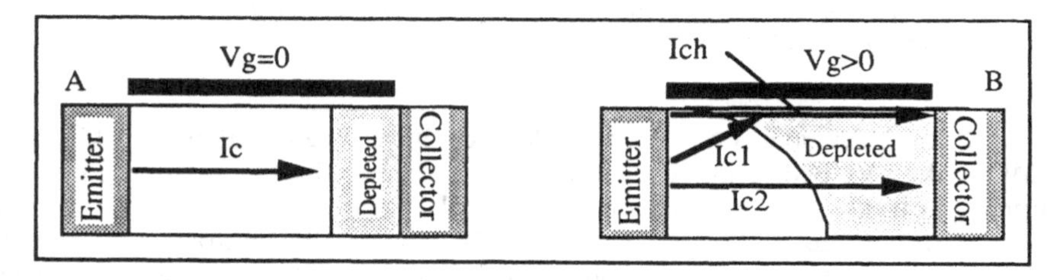

Figure 6.2.5: Emitter-to-collector current (I_c) flowing from emitter to collector in the lateral bipolar transistor with no applied gate bias (A), and the different current components flowing from source (emitter) to drain (collector) when a gate bias is applied to both gate and base: the MOS channel current (I_{ch}), a bipolar current component collected by the channel, (I_{c1}), and a bipolar current component flowing from emitter to collector (I_{c2}) (B).

The electrical characteristics of the VCBM device are presented in Figure 6.2.6. In this example, the threshold voltage of the MOS part of the device (0.5 V) is lower than that of the bipolar part of the device (≅0.7V). The subthreshold characteristics are similar to those of a MOSFET, and extra drive current is obtained once the bipolar turns on. The transconductance enhancement obtained by the "symbiotic" association is also quite clear from Figure 6.2.6, at least for gate voltages right above threshold. At higher gate voltages, the bipolar component of the current is reduced because of current crowding in the base and high-injection problems, and the total VCBM current is almost reduced to the MOS current component.

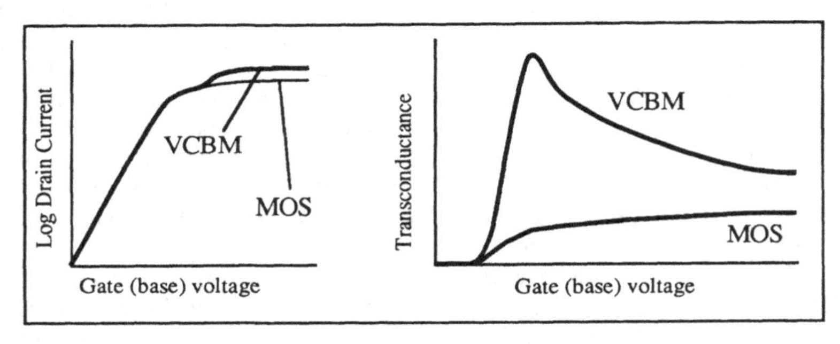

Figure 6.2.6: Comparison of the drain current (log scale) and the transconductance of a VCBM device and the associated MOS transistor (*i.e.*: the same device, without the body being connected to the gate). VDS=3 V in both cases.

Both NPN and PNP VCBM devices have been realized [6.11], and CMOS-like VCBM ring oscillators have been shown to operate with supply voltages of a fraction of a volt, with extremely low power consumption (gate delay of 4 ns, 0.7 μW power dissipation, and power-delay product of 3 fJ at a supply voltage of 0.7 V). Such a low power dissipation may render the device useful for body-implanted electronics.

6.2.3. Double-gate MOSFET

As seen in Chapter 2, the silicon-on-insulator MOSFET has two gates, the front gate and the back gate. Usually, the back gate oxide (the buried oxide) is much thicker than the front gate oxide (t_{ox1} =...20 nm..., while t_{ox2} = 400 ... 1000 nm or more), such that prohibitively high back-gate voltages have to be used for actively controlling a back inversion layer in the device. Furthermore, it is believed that the buried oxide might not be of sufficient quality to be used as a gate oxide. It is, however, possible to realize SOI MOSFETs with thin, gate-quality oxides at both the front and the back of the active silicon film [6.12, 6.13]. A schematic cross-section of a double-gate MOSFET is presented in Figure 6.2.7.

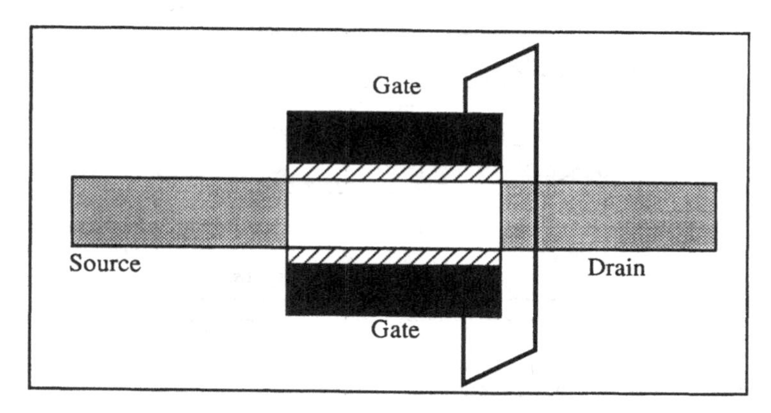

Figure 6.2.7: Double-gate MOSFET.

Such double-gate MOSFET can, for instance, be realized by etching away locally the buried oxide underneath the channel region of a device, such that a silicon "bridge" is formed. This bridge is supported by its extremities (which will later on become the source and the drain). A thermal gate oxide is then grown on the exposed silicon (on the top, bottom and edges of the free-standing parts of the silicon bridge). Polysilicon gate material is then deposited in an LPCVD reactor. The excellent step coverage of LPCVD polysilicon deposition allows for the gate material to cover the gate oxide everywhere, even underneath the active silicon bridge [6.13]. Such a device (gate-all-around MOSFET) is presented in Figures 6.2.8 and 6.2.9.

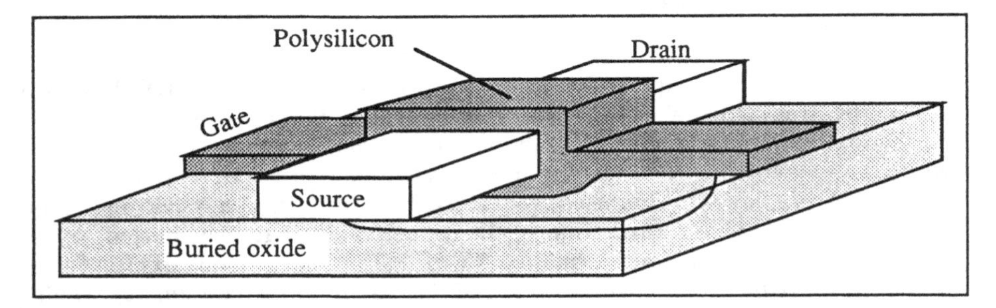

Figure 6.2.8: Bird's eye view of the "gate-all-around" MOSFET.

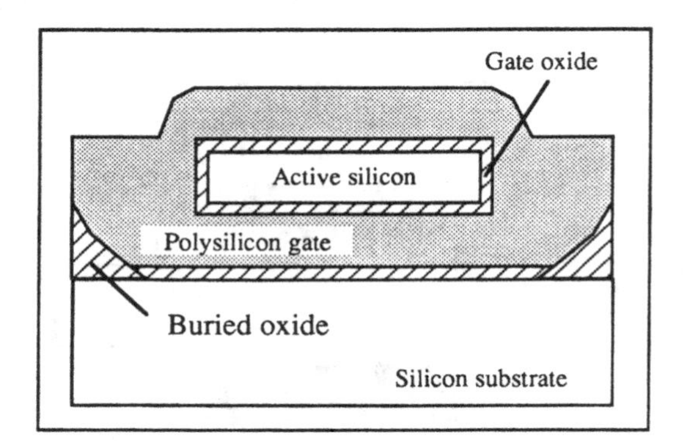

Figure 6.2.9: Cross section of the "gate-all-around" MOSFET, perpendicular to the direction of current flow.

Double-gate devices, where thin, gate-quality oxide is grown at both sides of a thin silicon film, exhibit near-ideal subthreshold characteristics and high transconductance. It is also predicted that they should have excellent short-channel behavior [6.14]. The presence of a gate both above and below the channel region allows for improved control of the potential between source and drain. As a consequence, subthreshold slopes of nearly 60 mV/decade are obtained, which proves excellent control of surface potential by the gate ($d\Phi_S \cong dV_G$). When the silicon film is thin enough and when the gate voltage is slightly larger than the threshold voltage, the variation of potential across the silicon film is very small [6.15], and the entire silicon film can be in inversion (*i.e.*: at any depth, *x*, in the silicon film, one has: $n(x) >> p(x)$) (Figure

6.2.10.A). Because of this property, the name "volume inversion MOSFET" has been proposed for such a device [6.16]. Right above threshold, the electron concentration is almost constant across the silicon film, and the contribution of the inversion within the bulk of the silicon film is important. It is also worthwhile noting that the carrier mobility in the volume is higher than the mobility in a surface channel, because of the absence of carrier scattering at the $Si\text{-}SiO_2$ interfaces.

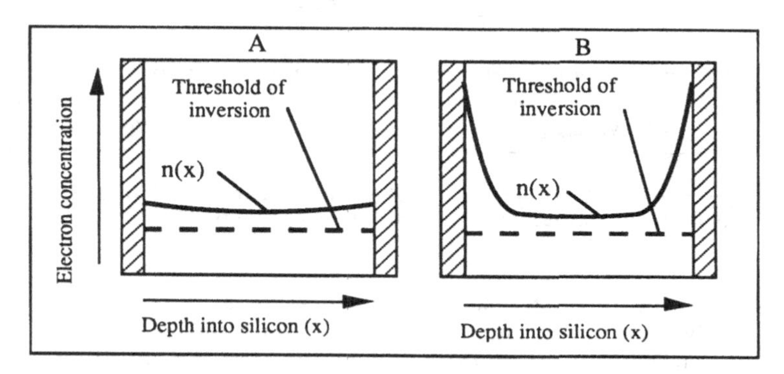

Figure 6.2.10: Electron concentration profiles in an n-channel volume-inversion MOSFET as a function of depth in the silicon film, right above threshold (A), and for $VG \gg V_{th}$ (B).

At higher gate voltage, however, the electron concentration does not increase much in the volume of the device, and two inversion "channels" form near the Si-SiO2 interfaces (Figure 6.2.10.B). As a consequence, the relative contribution of volume inversion to the total drain current becomes smaller than right above threshold. The apparent mobility also decreases rapidly as the conduction takes place predominantly in the inversion "channels". Figure 6.2.11 presents the measured transconductance, dI_D/dV_G, of a double-gate MOSFET as a function of gate voltage as well as that of a reference "single-gate" (conventional) MOSFET. Figure 6.2.11.A also presents the transconductance curve of the conventional device multiplied by two (to account for the presence of a top and a bottom channel) and shifted to the left by the difference of threshold voltages between the double-gate and the single-gate MOSFETs (such that comparison is made for same values of $V_G\text{-}V_{th}$). The increase of current drive capability of the double-gate device is represented by the shaded area and is attributed to the contribution of volume inversion.

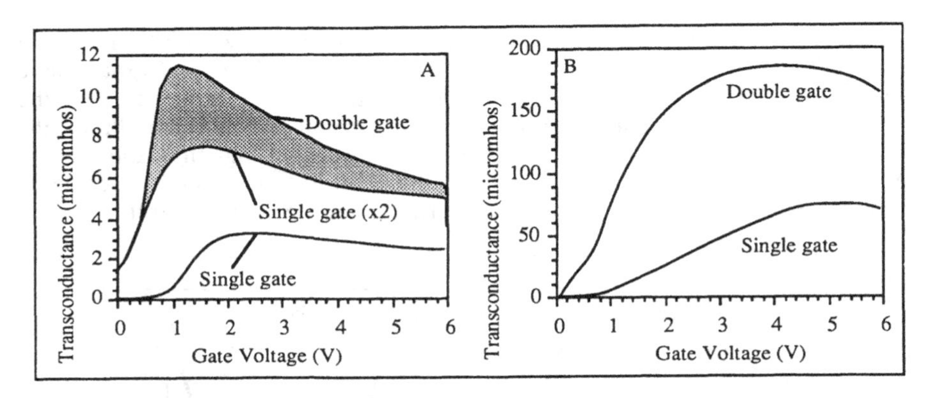

Figure 6.2.11: Transconductance of a "gate-all-around" MOSFET as a function of gate voltage for V_{DS}=100 mV (A) and V_{DS}=2 V (B).

Beside the high transconductance, such a device offers another potential advantage. The threshold voltage shift induced by the creation of oxide charges in a MOSFET submitted to an ionizing radiation environment (X-rays, gamma-rays) is roughly proportional to the square of the thickness of the oxides in contact with the silicon. Indeed, the amount of oxide charges, ΔQ_{ox}, is proportional to the oxide thickness, t_{ox}, and the threshold voltage shift, ΔV_{th}, is equal to $\Delta V_{th} = \Delta Q_{ox}.t_{ox}/\varepsilon_{si}$, such that $\Delta V_{th} \propto (t_{ox})^2$. In a conventional SOI MOSFET, the buried oxide (back-gate oxide) is quite thick (400...1000 nm) and, as a result, the threshold voltage shift induced by ionizing radiations in thin, fully-depleted SOI MOSFETs is quite dramatic [6.17]. If a double-gate device is used, on the other hand, one might expect much less device degradation, since the active channel area is in contact with only thin, thermally-grown (high-quality) oxide.

6.2.4. Bipolar transistors

Lateral bipolar transistors are unfortunately mostly known as parasitic elements in SOI MOSFETs. Good lateral bipolar devices can readily be fabricated using an SOI CMOS process. Current gains up to 70 and 40 have been demonstrated in NPN and PNP devices, respectively, for base widths (assumed to be equal to the gate length) of 0.45 μm

[6.18]. The maximum gain, however, peaks at relatively low collector current values, on the order of a microampere. This is due to both the high base resistance caused by the lateral base contact, as well as high current injection (high current densities are reached very quickly in thin films). The use of a top base contact can remedy the base resistance problem, and higher current levels can be reached [6.19]. The performance of this device is, however, limited by a rather high collector resistance, high injection, and recombination in the base region.

Vertical bipolar transistors can, of course, be made in thick SOI material [6.21-23]. Owing to the total dielectric isolation between the devices, complementary (*i.e.*: NPN and PNP) vertical bipolar transistors can even be integrated on a same chip [6.2]. Because of the need for making a buried collector layer, it may seem impossible to fabricate vertical bipolar transistors in thin SOI films. An original solution to this problem has, however, been proposed, and vertical NPN transistors have been realized in 400 nm-thick SOI films (and the use of even thinner films is quite possible) [6.20].

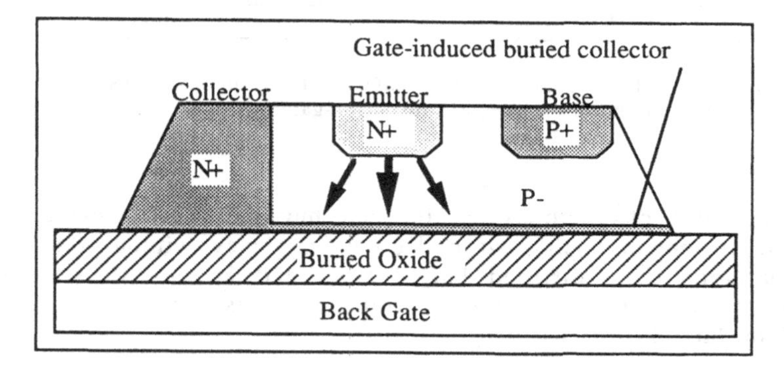

Figure 6.2.12: Vertical bipolar transistor with gate-induced buried collector.

The solution consists in fabricating a vertical bipolar device with no buried collector diffusion. The bottom of the P- intrinsic base is directly in contact with the buried oxide. If no back gate bias is applied, very few of the electrons injected by the emitter into the base can reach the lateral collector, and the current gain of the device is extremely small (β<<1, see Figure 6.2.13). When a positive bias is applied to the back gate, however, an inversion layer is induced at the silicon-buried oxide interface, at the bottom of the base, and acts as a buried collector

(Figures 6.2.12-13). Furthermore, the band bending in the vicinity of the bottom inversion layer attracts the electrons injected by the emitter into the base. As a result, collection efficiency is drastically increased, and the current gain can reach useful values (β=50 [6.20]).

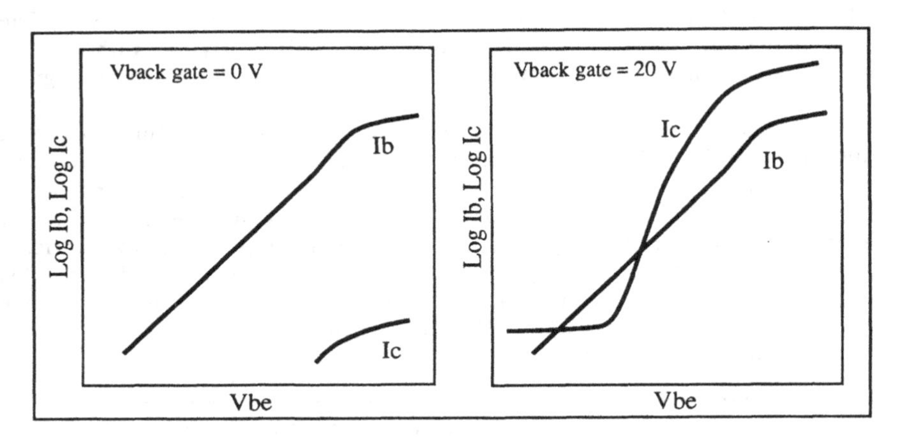

Figure 6.2.13: Gummel plot of SOI vertical bipolar transistor with gate-induced buried collector for back-gate voltages of 0 and 20 volts.

The use of a field-effect-induced buried collector makes it thus possible to obtain significant collection efficiency without the need for the formation of a diffused buried layer and an epitaxy step. It also solves the high-injection (high current densities) problems inherent to thin-film lateral devices.

6.2.5. Optical modulator

The presence of a buried layer underneath an active silicon layer makes it possible to use the SOI structure to realize optical modulators. Indeed, an SOI structure capped by an oxide and polysilicon can form a Fabry-Perot cavity [6.25]. Infrared laser light (λ=1.3 μm)can be introduced into the cavity by means of a single-mode fiber optics. The light bounces forth and back inside the cavity, partially reflected through, and partially reflected back. Resonance can be obtained by optimizing the thickness of the different film thicknesses of the multilayer structure. At resonance, the reflectance is minimized.

Forward biasing the PIN diode of the structure generates carriers inside the intrinsic (actually, N^-) region which modulates the phase of the laser light in the cavity and shifts the resonance of the Fabry-Perot cavity to convert the phase modulation into intensity modulation. At resonance, a 0.6% change of the refractive index in the "intrinsic" portion of the PIN diode can theoretically induce a 65% reduction of the reflectance.

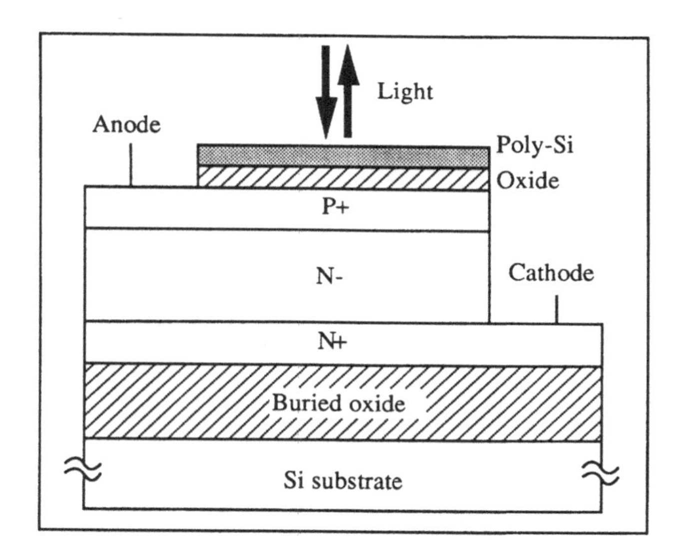

Figure 6.2.14: SOI infrared optical modulator.

Using such a structure, a 40% modulation depth of 1.3 μm infrared laser light has been demonstrated, the 40% optical output reduction being obtained for a 2.25 A/cm^2 current density in the PIN diode [6.25].

CHAPTER 7 - The SOI MOSFET Operating in a Harsh Environment

SOI MOSFETs present several properties which allow them to operate in harsh environments where bulk devices would typically fail from operating satisfactorily. These interesting properties of the SOI MOSFETs are due to the small volume of silicon in which the devices are made, to the small area of the source-body and drain-body junctions, and to the presence of a back gate. In this Chapter, we will describe the behavior of the SOI MOSFET operating in three cases of extreme environments: the exposure to radiations, to high temperatures and to low (cryogenic) temperatures.

7.1. Radiation environment

One of the major niche markets where SOI circuits and devices are currently employed is the aerospace/military market, because of the high hardness of SOI technology against transient radiation effects. The effect of radiation on an electronic devices depends on the type of radiation (neutrons, heavy particles, electromagnetic radiations,...) to which the devices are submitted. Unlike bipolar devices, MOSFETs are relatively insensitive to neutron irradiation (neutrons basically kill the carrier lifetime in silicon through inducing displacement of atoms within the crystal lattice). MOS devices are much more sensitive to the exposure to single-event upset (SEU), gamma-dot upset, and total-dose exposure than to neutrons. The effects created in bulk silicon MOSFETs by such radiation exposures are well documented and can be found in reference [7.1] for instance. The following Sections will compare the hardness of SOI and bulk devices to SEU, gamma-dot and total-dose irradiation conditions.

7.1.1. SEU

Single-event upset (SEU) is caused by the penetration of an energetic particle, such as an alpha particle or an heavy ion (cosmic ray) within a device. Indeed, when such a particle penetrates a reverse-biased junction, its depletion layer and the bulk silicon underneath it, it produces a track along which the silicon becomes ionized (electron-hole pairs are created) [7.2]. The presence of this track temporarily collapses the depletion layer and distorts the equipotential surfaces of the depletion layer electric field in the vicinity of the track. The distortion of the depletion layer is called "funnel" (Figure 7.1.1).

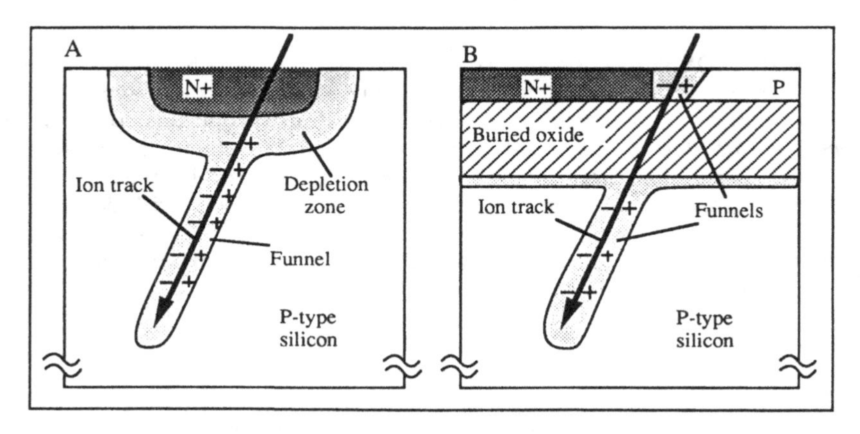

Figure 7.1.1: Penetration of bulk (A) and SOI (B) devices by an energetic ion.

The "funneling" of potential surfaces produces a large electric field such that the electron-hole pairs created along the track are separated. In a bulk device, the electrons are propelled back up the funnel into the depletion layer (Figure 7.1.1.A). The holes move downwards and create a substrate current. The electrons result in a large increase of the charge collected by the depletion layer, which can upset the logical state at which the node was. The duration of the collection of the electrons by the node is on the order of a few tens of picoseconds. The length of the track in the silicon is typically on the order of ten micrometers. In an SOI device, the impinging particle ionizes the silicon along its track as well. However, because of the presence of a buried insulator layer between the active silicon film and the substrate, none of the charges generated within the substrate can be collected by the junctions of the SOI devices. The only electrons which can be collected are those produced within the thin silicon film, the thickness of which is typically 150-300 nm in rad-hard applications. The ratio of the lengths of the tracks along which the electrons are collected gives a first-order approximation of the advantages of SOI over bulk in terms of SEU hardness (*e.g.*: $\frac{200\ nm}{10\ \mu m}$ = 50 in the case of a 200 nm-thick SOI device).

The magnitude of SEU is expressed in linear energy transfer (LET) units. It is defined by the following relationship: LET = $\frac{1}{m_v}\frac{dW}{dx}$, where x is the linear distance along the particle track, dW is the energy lost by the particle and absorbed by the silicon, and m_v is the volumic mass of silicon. The LET is usually expressed in MeV·cm^2/mg. The number of electrons or holes created by a single-event upset is given by: $\frac{dN}{dx}=\frac{dP}{dx}=\frac{m_v}{w}$ LET, where w is the energy needed to create an electron-hole pair [7.3]. As an illustration, a single carbon ion with an energy of 1 GeV (LET ≅ 0.24 MeV·cm^2/mg) produces 1.5x10^4 pairs/µm in silicon. If it hits a bulk silicon device, it produces 1.5x10^5 electron-hole pairs (≅0.3 pC) along the 10 µm-track. If the electrons migrate towards the node within a time scale of 10-100 ps, an SEU current spike of 1-10 mA is then created. Because of the reduced length of the track along which the electrons are collected in SOI, the SEU current spike will be a factor ...50... smaller in an SOI device than in a bulk device.

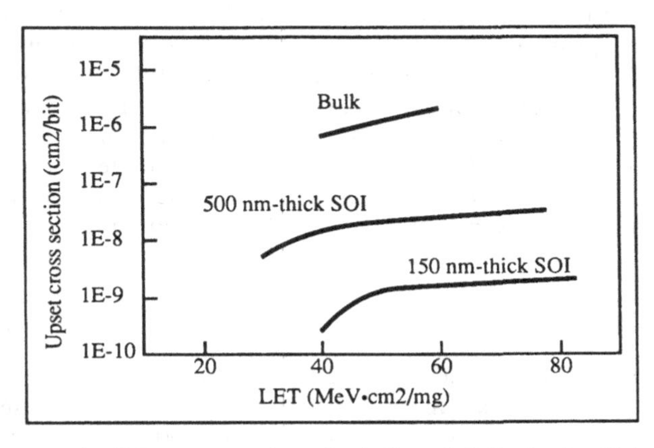

Figure 7.1.2: SEU cross sections for bulk and SOI circuits [7.4, 7.5].

The single-event upset cross-sections of several bulk and SOI circuits are presented in Figure 7.1.2 [7.4, 7.5]. The SEU cross section is the equivalent area of a logical bit (expressed in cm^2/bit). The smaller this cross-section, the less the devices are sensitive to irradiation. One can see that 150 nm-thick SOI devices are approximately 10 times less sensitive than thicker (500 nm) devices which, in turn, are about 50 times harder than bulk CMOS devices.

It is worth noting that the generation of a funnel in the substrate underneath the buried oxide of an SOI structure can also influence the characteristics of a device submitted to SEU. Indeed, if the substrate (back-gate) bias is such that the substrate under the buried oxide is depleted, the electrons created along the track will move upwards towards the buried oxide (Figure 7.1.1.B). These electrons immediately induce a positive mirror charge in the upper silicon layer. As a result, electrons are injected into the device by the external circuit to restore equilibrium, and the occurrence of a transient current is measured [7.5]. This effect is not observed if the surface of the substrate underneath the buried oxide is either inverted or accumulated.

The photocurrent - or "ionocurrent" - generated by the impact of a particle in an SOI MOSFET can be amplified by the parasitic lateral bipolar transistor present in the device [7.6, 7.7]. Indeed, the hole current I_B created within the body of an SOI MOSFET serves also as base current for the parasitic lateral NPN bipolar transistor (n-channel device case). In response to the base current pulse induced by the particle, a collector current $I_C=\beta\ I_B$ is produced. This current adds to the current pulse caused by the SEU-induced electrons collected by the drain. Therefore, the presence of any bipolar transistor action, even with low gain ($\beta<1$) contributes to enhance the transient current produced by SEU [7.7], but the problem is, of course, more pronounced in short-channel devices where β is large. The solutions proposed to remedy this problem are the use of a body contact through which part of the base current can be evacuated, and the use of SOI material with poor carrier lifetime [7.6]. Circuit solutions (in contrast to device or technology solutions) include the increase of RC time constant of the different circuit nodes, but this technique presents the drawback of degrading the circuit speed performances [7.6].

7.1.2. Total dose

Total-dose effects are caused by the cumulative exposure to ionizing radiation such as X-rays and gamma rays. The dose unit is the rad(Si) which is defined by the deposition of 100 erg of radiation energy per gram of silicon. The rad(Si) is a CGS unit, but there exists a MKS unit for dose called the gray (Gy). One Gy is defined as the deposition of 1 joule of radiation energy per kilogram of matter. The equivalence between the two units is straightforward: 1 Gy = 100 rads. The number of electron-hole pairs which are generated is related to the energy dW absorbed by volume unit dv of the material: $\frac{dN}{dv}=\frac{dP}{dv}=\frac{1}{w}\frac{dW}{dv}$ where w is the effective energy needed to produce a pair in silicon (w=3.6 eV). The relationship between the dose, D, and the number of pairs generated, is given by: $\frac{dN}{dv} = \frac{dP}{dv} = \frac{m_v}{w} D$ [7.3]. It is generally admitted that 1 rad(Si) generates $4x10^{13}$ pairs cm^{-3} in silicon and $7.6x10^{12}$ pairs cm^{-3} in SiO_2. To get an idea of the magnitude of the levels of irradiation to which devices can be submitted, one can mention the following numbers: a medical or dental X-ray corresponds to 0.3 rad(Si). The human being becomes sick after being exposed to 100 rad(Si) and falls in an instant coma if submitted to 10 krad(Si). During their lifetime, satellites orbiting around the earth receive total doses ranging between 10 krad(Si) and 1 Mrad(Si), depending on the orbit parameters. Interplanetary spacecrafts and some electronics in nuclear reactors can be exposed to doses in excess of 10 Mrad(Si) [7.8]. Some SOI circuits have been tested at doses up to 500 Mrad(Si) [7.9].

The main effect caused by total dose in MOS devices is the generation of charges in the oxides and surface states at the Si/SiO_2 interfaces. If the rate of energy deposition is high, a sufficient amount of electron-hole pairs can be created in the silicon to produce photocurrents. This case, where dD/dt is high, will be considered in the next Section (gamma-dot effects). Ionizing electromagnetic radiations such as X-rays and gamma rays (produced *e.g.* by a ^{60}Co source) create electron-hole pairs in silicon dioxide. Electrons are fairly mobile in SiO_2, even at room temperature, and can move rapidly outside the oxide (towards a positively biased gate electrode, in the case of a gate oxide, for instance). Holes, on the other hand, remain trapped within the oxide and contribute to the creation of a positive oxide charge, Q_{ox}. If we take the example of a gate oxide, the charge Q_{ox} is proportional to the thickness of the oxide, t_{ox}, and the resulting threshold voltage shift is, therefore, proportional to t_{ox}^2 since $\Delta V_{th} = -\frac{Q_{ox}\, t_{ox}}{\varepsilon_{ox}}$. The relationship between the threshold voltage shift and the dose can be written [7.3]:

$$\Delta V_{th} = -\alpha \frac{q\, m_v}{w\, \varepsilon_{ox}} t_{ox}^2 D$$

where w is the effective energy needed to produce a pair in oxide (18 eV), m_v is the volumic mass of oxide and where the factor α accounts for the fact that only a fraction of the charges become trapped within the oxide. α is a technology-dependent parameter. Typical values of α are 0.15 for normal, unhardened oxides and 0.05 or less for rad-hard oxides. SIMOX buried oxides have an α of approximately 0.05 [7.3]. The physics of the irradiation of devices is, unfortunately, much more complex than what is described in the few equations above. For example, the rate at which the energy is deposited into the device has considerable influence. Dose irradiation also creates surface states at the $Si\text{-}SiO_2$ interfaces. Contrarily to oxide charges, which decrease the threshold voltage in n-channel devices, interface states increase it. Under high-dose irradiation conditions, above a threshold which depends on the technology used to fabricate the devices, more interface states than oxide charges are produced, and the threshold voltage starts to increase (after having decreased for a while under lower dose exposure) [7.10]. In addition, oxide charges anneal out with time and even contribute to the creation of additional interface traps. The latter effect is called "rebound" and illustrated in Figure 7.1.3 [7.11]. This effect is, of course, temperature dependent.

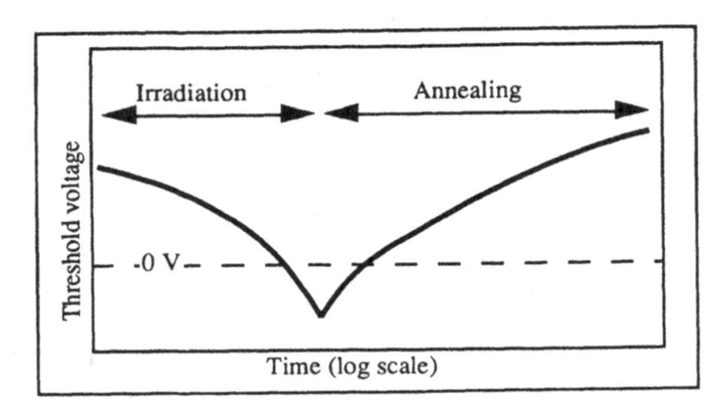

Figure 7.1.3: "Rebound" effect in an SOI n-channel MOSFET exposed to dose irradiation. Annealing occurs after irradiation has stopped.

Furthermore, the dose-induced effects are highly dependent on the front- and back-gate bias of the device. Worst-case conditions correspond to positive gate biases (which push the holes in the oxide towards the Si-SiO_2 interfaces). In p-channel transistors, the generation of both positive charges and interface states cause an increase of the absolute value of the threshold voltage (*i.e.*: it becomes more negative). Bias gate conditions are also different (front- and back-gate biases are usually zero or negative).

The SOI MOSFET, with its many Si-SiO_2 interfaces (gate oxide, buried oxide and edge (field) oxide), is quite sensitive to total-dose exposure, unless special hardening techniques are used. While SOI gate oxide hardening techniques are similar to those employed in bulk (*e.g.*: low-temperature oxide growth), the techniques for prevention of back and edge leakage are specific to SOI. The classical solution for avoiding the formation of an inversion layer at the bottom of the silicon film in n-channel devices consists in increasing the threshold voltage of the back transistor by means of a back boron impurity implant. In some instances, use of a back-gate bias can be made to compensate for the radiation-induced generation of positive charges in the buried oxide. The creation of a peak of boron doping at the back interface implies the use of partially-depleted devices. Indeed, fully depleted are usually too thin to allow one to realize such an implant (see Section 4.3). In addition, the presence of charges in the buried oxide and interfaces states at the silicon - buried oxide interface causes front threshold voltage shifts and degrades the performance of the devices if the front and back interfaces are electrically coupled, which is the case in thin-film, fully-depleted devices. SOI circuits (non-fully depleted) made in a silicon film thickness as low as 150 nm and capable of withstanding doses up to 300 Mrad(Si) have, however, been reported [7.5]. It has also been shown that, in some instances, the back-channel leakage current taking place in irradiated devices made in SOI can also be traced back to the presence of metal impurities in the silicon film [7.13].

The control of edge leakage currents in n-channel devices poses an even more severe problem. Indeed, the parasitic MOS transistors can be quite sensitive to dose irradiation because of the relatively thick oxides (such as LOCOS or oxidized mesa oxides) which are in contact with the edges of the silicon active areas (see Section 4.2). An example of radiation-induced edge-leakage current is presented in Figure 7.1.4 [7.5].

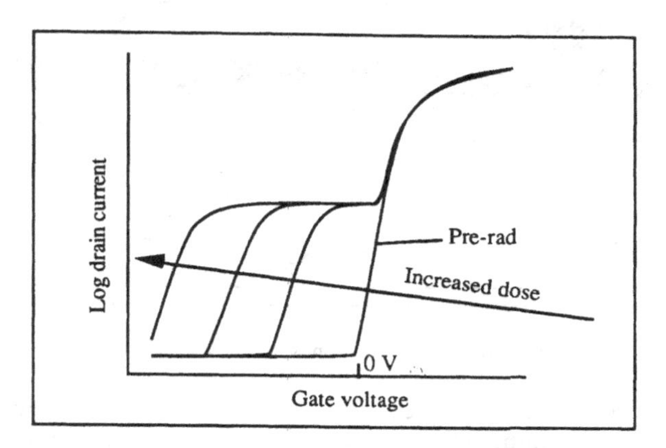

Figure 7.1.4: Edge leakage current caused by dose irradiation.

Edge leakage problems can be eliminated through optimization of the sidewall doping and field oxide growth processes [7.6] or through the use of edgeless devices or transistors where a P^+ diffusion interrupt edge leakage paths between the N^+ source and drain diffusions (see Section 4.6). There is usually no edge leakage problem in p-channel devices.

7.1.3. Dose-rate

Dose-rate (D') effects take place when a large dose of electromagnetic energy (X-rays, gamma rays) is deposited within a short time interval by such events as *e.g.* solar flares or nuclear explosions. Dose-rate effects are also often referred to as gamma-dot effects. The dose-rate unit is the rad(Si).s^{-1}. As mentioned in the previous Section, 1 rad(Si) generates approximately $4x10^{13}$ pairs cm^{-3} in silicon. If the energy is absorbed by the silicon in a short time, a significant amount of photocurrent can be generated within the depletion zones of the devices. The generated photocurrent is expressed by: $I_{ph}=q \cdot V_{depl} \cdot g \cdot D'(t)$ where q is the electron charge, V_{depl} is the volume of the depletion zone, and g is the carrier generation constant in silicon, which is equal to $4.2x10^{13}$ hole-electron pairs cm^{-3} rad(Si)$^{-1}$ [7.14]. The major difference between bulk and SOI devices resides into the much smaller depletion volume, V_{depl}, found in SOI devices (Figure 7.1.5). As a result, the dose-rate-induced photocurrent is significantly smaller in SOI transistors than in bulk devices.

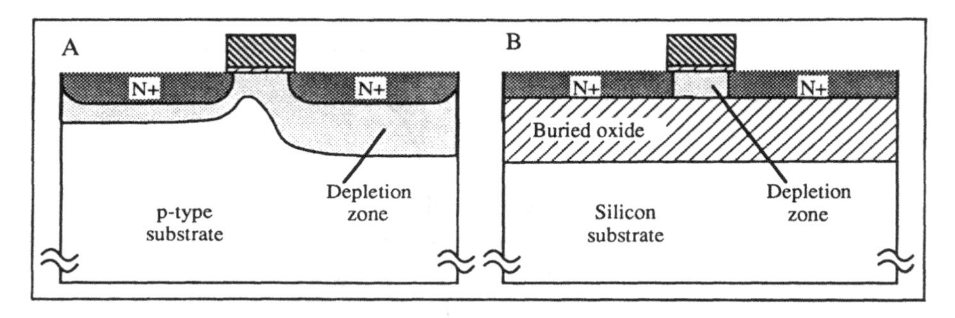

Figure 7.1.5: Volume of the depletion zones in bulk (A) and SOI (B) MOSFETs.

As in the case of SEU, the photocurrent generated in an SOI MOSFET can be amplified by the parasitic lateral bipolar transistor present in the device [7.4, 7.6, 7.7]. Indeed, the hole current created within the body of an SOI MOSFET serves also as base current for the parasitic lateral NPN bipolar transistor (n-channel device case). In response to the base current pulse I_B induced by the particle, a collector current $I_C=\beta\ I_B$ is produced. This current adds to the photocurrent caused by the gamma-dot event. Therefore, the presence of any bipolar transistor action, even with low gain ($\beta<1$) contributes to enhance the transient current produced by a gamma-dot event [7.7] but the problem is, of course, more pronounced in short-channel devices where β is large. The solutions proposed to remedy this problem are the use of a body contact through which part of the base photocurrent can be evacuated, and that of SOI material with poor carrier lifetime [7.4, 7.6].

7.2. High-temperature operation

Some applications (well logging, avionics, automotive,...) require electronic circuits capable of operating at temperatures up to 300°C. SOI devices and circuits present three advantages in this field over their bulk counterparts: the absence of thermally-activated latch-up, reduced leakage currents and, in thin-film, fully-depleted devices, a smaller variation of threshold voltage with temperature. The absence of latch-up in SOI is quite obvious and will not be discussed here, although it is an important argument for the use of SOI circuits for high-temperature applications.

7.2.1. Leakage currents

The increase of junction leakage current is one of the main cause of failure in circuits operating at high temperature. The junction leakage current is, of course, proportional to the area of the junctions. The strongest leakage component found in bulk CMOS operating at high temperatures originates from the "huge" well junctions [7.15]. These cause loss of functionality in CMOS circuits at temperatures above ...240°C... , while the same SOI CMOS circuits (4k SRAMs in Ref. [7.15]) were still operating at 300°C. Operation of SOI CMOS ring oscillators at temperatures up to 500°C has even been demonstrated [7.16].

The leakage current of a N^+-P reverse-biased junction is expressed by [7.20]:

$$I_{leak} = q\,A\left(\frac{D_n}{\tau_n}\right)^{1/2}\frac{n_i^2}{N_a} + q\,A\frac{n_i\,W}{\tau_e}$$

where q is the electron charge, A is the junction area, D_n is the electron diffusion coefficient, τ_n is the electron lifetime in p-type neutral silicon, n_i is the intrinsic carrier concentration, N_a is the doping concentration in the p-type material, W is the depletion width, and $\tau_e=(\tau_n+\tau_p)/2$ is the effective lifetime related to the thermal generation process in the depletion region [7.20]. The temperature dependence of n_i is given by equation 7.3. The first term in the above equation (diffusion component) is proportional to n_i^2, and the second term (generation component) is proportional to n_i. It is experimentally observed [7.17] that the leakage of SOI MOSFETs and reverse-biased diodes varies as n_i below a temperature of ...100-150°C... and as n_i^2 above that temperature. It is worth noting that A, the area of the junction, and $A{\cdot}W$, the volume of the space-charge region associated with the diode, are both much smaller in SOI transistors than in bulk devices. Therefore, extremely low leakage current can be observed in SOI devices at room temperature (*e.g.*: 64k SRAMs with a 10 nA standby current (V_{DD}=5 V) have been realized [7.21]), as well as at high temperature [7.15]. The leakage current (V_{DS} = 5 V) of bulk and SOI n-channel transistors with same geometries are compared in Figure 7.2.1, where the slightly higher leakage in the SOI device at room temperature is attributed to a relatively poor value of τ_e in the measured device. At high temperature, however, where diffusion current dominates, the leakage current is three orders of magnitude lower in the SOI devices than in the bulk device, owing to a much smaller junction area.

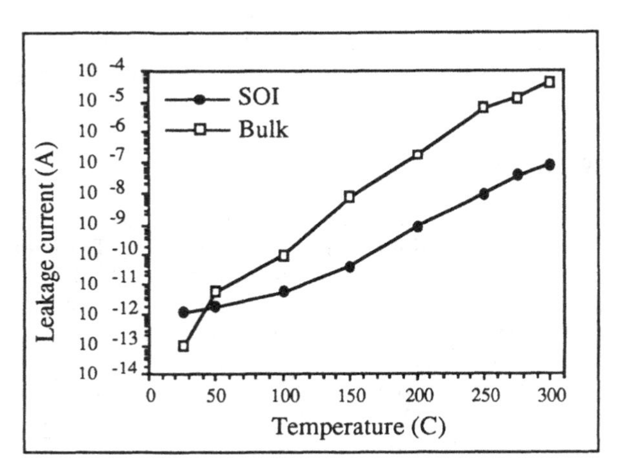

Figure 7.2.1: Leakage current in bulk and SOI n-channel transistors of same geometries as a function of temperature.

7.2.2. Threshold voltage

The threshold voltage of an n-channel MOSFET (bulk or SOI) is given by:

$$V_{th} \cong \Phi_{MS} + 2\Phi_F - Q_{ox}/C_{ox} - Q_{depl}/C_{ox} \qquad (7.1)$$

where Φ_{MS}, Φ_F, Q_{ox}, Q_{depl} and C_{ox} are the metal-semiconductor work function difference, the Fermi potential, the charge density in the gate oxide, the depletion charge controlled by the gate, and the gate oxide capacitance, respectively. Equation (7.1) is valid for partially depleted devices where Q_{depl} is replaced by q N_a x_{dmax}
x_{dmax} is the maximum depletion depth allowed for the doping concentration, N_a, under consideration, and is equal to $\sqrt{\frac{4\,\varepsilon_{si}\,\Phi_F}{q\,N_a}}$. In thin, fully depleted films, Q_{depl} depends on back-gate bias, and has a value which can range between q N_a t_{si} and q N_a $t_{si}/2$, q being the

electron charge, k the Boltzmann constant, N_a the doping concentration and t_{si} the silicon film thickness. If an N^+ polysilicon gate is assumed, the value of the work function difference is given by:

$$\Phi_{MS} = -\frac{E_g}{2} - \Phi_F \quad \text{with } \Phi_F = \frac{kT}{q}\ln(N_a/n_i) \tag{7.2}$$

with E_g, T and n_i being the silicon bandgap, the temperature and the silicon intrinsic carrier concentration, respectively. The temperature dependence of the intrinsic carrier concentration, n_i, is given by [7.19]:

$$n_i = 3.9\ 10^{16}\ T^{3/2}\ e^{-(Eg/2kT)} \tag{7.3}$$

The temperature dependence of the threshold voltage can be obtained from equations (7.1) and (7.2). For simplification, we will assume that Q_{ox} shows no temperature dependence over the temperature range under consideration and we will not take into account the presence of surface states at the Si-SiO_2 interfaces. E_g is assumed to be independent of temperature over the temperature range under consideration (20 - 250°C). Its actual variation is 0.3% [7.19]. In the case of a bulk or a thick-film SOI device, one obtains:

$$\frac{dV_{th}}{dT} = \frac{d\Phi_F}{dT}\left[1+\frac{q}{C_{ox}}\sqrt{\frac{\varepsilon_{si}\,N_a}{k\ T\ \ln(N_a/ni)}}\ \right] \tag{7.4}$$

with $$\frac{d\Phi_F}{dT} = 8.63\ 10^{-5}\ [\ln(Na) - 38.2 - \frac{3}{2}\{1+\ln(T)\}] \tag{7.5}$$

In the case of a thin-film, fully-depleted device, the depletion charge, Q_{depl}, is equal to q N_a t_{si}/n, where t_{si} is the silicon film thickness, which is independent of temperature, and the value of *n* is ranging between 1 and 2, depending on oxide charge and back-gate bias conditions. If one assumes that *n* is independent of temperature, the following dependence for the thin-film device is obtained:

$$\frac{dV_{th}}{dT} = \frac{d\Phi_F}{dT} \tag{7.6}$$

It can be seen by mere comparison of (7.4) and (7.6), that dV_{th}/dT is smaller in thin SOI devices than in bulk or thick SOI devices. The ratio of threshold voltage variation in bulk devices to that in thin SOI devices is given by the bracket term of (7.4), and is typically ranging from 2 to 3, depending on the gate oxide thickness and the channel doping concentration. The right-hand term of the bracketed expression of (7.4) is indeed caused by the temperature-induced variation of the depletion zone depth, variation which is obviously nonexistent in fully depleted devices. Typical values for dV_{th}/dT range between -0.7 and -0.8 mV/K in thin SOI MOSFETs. Much larger values of dV_{th}/dT are observed in bulk and thick-film SOI devices, such as presented in table 7.2.1 [7.18].

	SOI, T=25°C	Bulk, T=25°C	SOI, T=200°C	Bulk, T=200°C
t_{ox}=19 nm N_a=1.6 10^{17} cm^{-3}	-0.74	-1.87	-0.8	-2.3
t_{ox}=32.5 nm N_a=1.2 10^{17} cm^{-3}	-0.76	-2.42	-0.82	-3.05

Table 7.2.1: Calculated temperature dependence of threshold voltage on temperature, dV_{th}/dT, (in mV/K), in bulk and thin (t_{si}=100 nm) SOI MOSFETs. The data are presented for two temperatures (25 and 200°C) and for the two sets of technological parameters (t_{ox} and N_a).

When temperature increases in thin-film devices, however, the intrinsic carrier concentration increases, and Φ_F decreases. This gives rise to a decrease of the maximum depletion width, in such a manner that the devices are no longer fully depleted above a given critical temperature, T_k, where x_{dmax} becomes smaller than $t_{si,eff}$, where $t_{si,eff}$ is the effective electrical film thickness "seen" from the top gate, *i.e.* the silicon film thickness minus the backside depletion zone generated by both the interface oxide charges and the influence of the back gate. Above T_k, the transistor behaves as a thick-film SOI MOSFET, since the film is no longer fully depleted, and the temperature dependence of the threshold voltage becomes similar to that observed in bulk devices. dV_{th}/dT is then governed by equation (7.4) (Figure 7.2.2) [7.18].

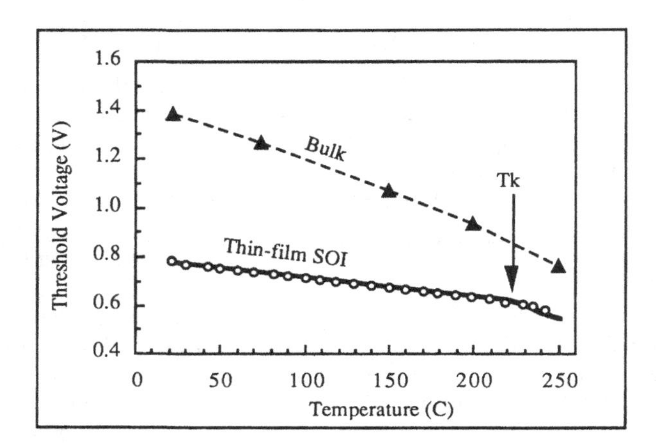

Figure 7.2.2: Variation of the threshold voltage of bulk and thin-film SOI n-channel MOSFETs with temperature [7.18].

7.3. Low-temperature operation

Low temperature (*e.g.*: 4.2 K (liquid helium) or 77 K (liquid nitrogen)) is not usually considered as a harsh environment as such, but low temperature is certainly a non-standard operating condition and has, therefore, been included in this Chapter. The effects of low-temperature operation in bulk MOSFETs are a shift of the threshold voltage (compared to the room temperature value), carrier freeze-out (and, as a consequence, the occurrence of kink effect), and improvement of the subthreshold slope. Due to reduced phonon scattering, the mobility is higher at low temperature than at room temperature. We will now discuss these low-temperature effects in SOI MOSFETs [7.22-7.25].

7.3.1. Threshold voltage

The value of the threshold voltage at can be obtained by extrapolating the expressions developed in Section 7.2 to low temperatures. Using equation (7.4) and taking the influence of interface traps into account, one obtains the following expression:

$$\frac{dV_{th}}{dT} = \frac{d\Phi_F}{dT}\left(1 + \alpha\,\frac{q}{C_{ox1}}\sqrt{\frac{\varepsilon_{si}\,N_a}{k\,T\,\ln(N_a/ni)}} + \frac{qN_{it}}{C_{ox1}}\right) \qquad (7.7)$$

where $\alpha=1$ in a partially-depleted device, and $\alpha=0$ in the fully-depleted case. N_{it} is the areal density of interface states. Electrical parameters being more sensitive to the presence of interface traps at low temperatures than at room or high temperatures, N_{it} cannot be neglected in the present low-temperature analysis. When temperature is decreased, fully-depleted devices remain fully depleted and exhibit low dV_{th}/dT values. Partially-depleted devices become fully depleted below the critical temperature, T_k, where x_{dmax} becomes equal to the silicon film thickness (*e.g.*: around 175 K for 200 nm-thick devices [7.24-7.25]), and dV_{th}/dT values of -1.9 mV/K and -0.6 mV/K are observed above and below T_k, respectively [7.25].

7.3.2. Kink effect

At low temperatures, kink effect is observed not only in thick-film SOI n-channel MOSFETs, but in bulk devices as well. Indeed, carrier freeze-out increases the resistivity of the silicon substrate to such an extent that it practically behaves like an insulator, and a kink is observed in the output characteristics of the devices, much like in a conventional partially-depleted SOI MOSFETs without a substrate contact [7.26]. Partially-depleted n-channel SOI devices, even with a body contact, show a kink under low-temperature operating conditions

[7.24-7.25]. Thin-film, fully-depleted devices show kink-free operation at low temperatures, probably for the same reason as they do at room temperature (the body-source potential barrier is very small) [7.25], but a kink is observed if a body contact is provided (which probably increases the body-source potential barrier [7.24].

7.3.3. Subthreshold slope

Using expressions (5.4.9) $S=\frac{kT}{q}\ln(10)\left(1+\frac{C_D+C_{it}}{C_{ox1}}\right)$ for a partially-depleted device and (5.4.17) $S\cong\frac{kT}{q}\ln(10)\left(1+\frac{C_{it}}{C_{ox1}}\right)$ for a fully-depleted device, one finds that the subthreshold slope varies almost linearly with temperature. The capacitance associated with the interface traps ($C_{it}=q\,N_{it}$) must be included in the above expressions because their influence on the electrical properties of the devices is more pronounced at low temperature than at room temperature. Experimental results, however, show a sublinear decrease of the subthreshold slope with decreased temperature. This is due to a strong inhomogeneity of the energy profile of the interface states in the band gap. At low temperatures the Fermi level approaches the band edges in such a way that the effective density of states and, therefore, C_{it}, increases rapidly. Subthreshold slopes of 23 mV/decade are observed in fully-depleted devices at 9 K.

7.3.4. Mobility

Mobility increases at low temperatures because of reduced phonon scattering. This well-known phenomenon is valid for SOI devices as well as for bulk MOSFETs. Mobilities up to 3000 $cm^2/V.s$ are observed in fully-depleted n-channel devices at 10 K. An interesting phenomenon appears in thin-film devices at low temperatures: after exhibiting a normal behavior from room temperature to $\cong$100 K, the mobility exhibits a sudden increase when the temperature is further reduced (μ_n = 2000 $cm^2/V.s$ at T= 80 K and 2600 $cm^2/V.s$ at T= 70 K). This has been attributed to a stronger coupling between the front and back interfaces at very low temperatures, which causes the potential in the entire silicon film to increase with gate voltage. Therefore, the distribution of minority carriers (electrons in an n-channel device) can now extend into the volume of the silicon film, and volume inversion is obtained. As a result, conduction takes place not only in the inversion channel, but also within the bulk of the silicon film. The carriers flowing in the volume have a much higher mobility that those confined within the inversion channel. As a result, a net apparent mobility increase is observed [7.25].

CHAPTER 8 - SOI Circuits

8.1. Rad-hard and high-temperature circuits

One of the major niche markets where SOI circuits and devices are currently used is the aerospace/military market, where radiation hardness is a key issue. High-level radiation hardness is also a prerequisite for circuits used in core instrumentation for nuclear power reactors (a containment accident may lead to more than 100 Mrad(Si) of dose exposure), nuclear well logging, control electronics in spacecraft nuclear reactors, reaction chamber instrumentation for particle colliders, and tokamak fusion reactor instrumentation. In some cases, these high-level irradiation environments may be associated with high temperature conditions [8.1]. Under these conditions, SOI has an advantage over bulk CMOS devices, since the reduced active silicon volume and the full dielectric isolation gives rise to lower leakage currents in SOI than in bulk. As we have seen in Section 7.1, SOI MOSFETs are inherently harder against SEU and gamma-dot than their bulk counterparts. SOI circuits are also free of latchup problems, which can be triggered by SEU or gamma-dot photocurrents in bulk CMOS circuits. Upon SEU or gamma-dot exposure SOI circuits such as 4k x 1 SIMOX SRAMs show bit error rates comparable to SOS but better than those obtained in bulk CMOS [8.2]. The main reason for not obtaining performances superior to those of SOS (a smaller silicon film thickness can indeed be utilized in SIMOX than in SOS) is the amplification of the SEU-induced photocurrent by the parasitic lateral bipolar transistor present in the SIMOX devices, where the minority carrier lifetime is high, compared to that of SOS material. Modern SIMOX memories (64k SRAMs) offer SEU error rates of 10^{-9} errors/bit-day (worst-case geosynchronous orbit error rate) and retain functionality at dose-rate exposures up to 10^{11} rad(Si)/sec [8.3]. These numbers represent an hardness improvement by a factor ≅100 over bulk silicon.

Total-dose hardening, on the other hand, does not come for free but can be obtained using an optimized fabrication process and special transistor design [8.4]. The degradation phenomena generated by dose exposure on CMOS circuits are a degradation of the response time (*e.g.* an increase of the access time in SRAMs) and the increase of power consumption. These are mostly due to the threshold voltage shift and the increase of leakage current in MOSFETs upon irradiation. Most rad-hard 16k and 64k CMOS SOI SRAMs are designed to remain functional after exposures to doses up to 1 Mrad(Si) [8.5-8.7, 8.10], but recent parts show 10 Mrad(SiO_2) hardness levels [8.8] (1 rad(SiO_2) = 0. 56 rad(Si) [8.9]). The highest total-dose hardness results reported to date for large circuits mention the operation of 29101 16-bit microprocessors at doses up to 100 Mrad(SiO_2) [8.4]. These circuits exhibit a 30% and 100% increase of worst-case propagation delay after 10 and 30 Mrad(SiO_2) exposure, respectively, and some parts are still functional after 100 Mrad(SiO_2) irradiation. Smaller CMOS/SIMOX circuits (frequency dividers) can show no propagation delay degradation at doses up to 1 Mrad(Si) and can even survive doses of 0.5 Grad(Si) [8.4, 8.10]. Figure 8.1.1 presents the static current consumption and the gate propagation delay of divide-by-16 CMOS/SIMOX circuits as a function of ^{60}Co dose exposure (dose-rate=2.8 Mrad(Si)/hour) [8.4].

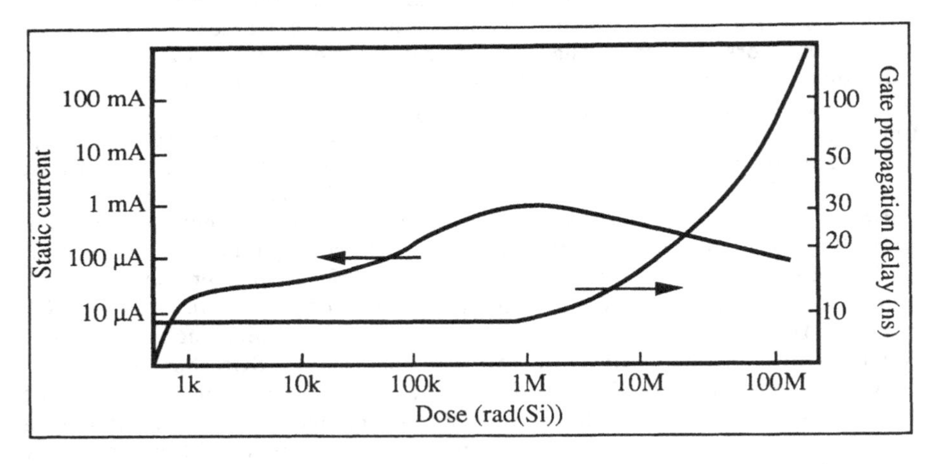

Figure 8.1.1: Total static current consumption and gate propagation delay in a divide-by-16 circuit realized using a CMOS/SIMOX hardened process.

High temperatures represent another form of harsh environment where SOI circuits present clear advantages over bulk devices. As presented in Section 7.2, SOI devices exhibit much lower leakage current than bulk devices at high temperature, owing to the much smaller area of source and drain SOI junctions. The absence of well junctions in SOI contributes even more to the difference of leakage current and power consumption between SOI and bulk CMOS circuits operating at high temperature. Excess power consumption is indeed a major source of failure in bulk CMOS circuits at high temperature, and 250°C seems to be an upper limit for bulk CMOS circuit operation [8.11]. On the other hand, 4k SOI CMOS SRAMs operating at 300°C have been reported [8.11]. Operation of SOI CMOS ring oscillators at temperatures up to 500°C has been demonstrated, and the loss of functionality during operation at these temperatures has been identified as a failure of the metallization system (aluminum), and not as a failure of the MOS devices [8.12].

8.2. VLSI and high-speed CMOS circuits

The area where SOI CMOS technology offers the largest potential for commercial success is submicron VLSI/ULSI. Indeed, as we have seen in Chapters 1, 4 and 5, thin-film SOI devices scale much better than bulk devices. Thin-film, fully depleted SOI MOSFETs and circuits present the following advantages over their bulk counterparts:

absence of latchup
higher soft-error immunity
higher transconductance
less processing yield hazards
reduced electric fields
reduced parasitic capacitances
reduced short-channel effects
sharper subthreshold slope
shorter and easier CMOS processing

On the other hand, two main issues have to be addressed before SOI can be considered as a serious competitor to bulk silicon for future deep-submicron CMOS applications. The first of these is the reduced drain breakdown voltage and the snapback problem caused by the parasitic bipolar transistor. The solution to these problems will probably consist into thorough source and drain engineering. Indeed, the LDD (lightly doped drain) structure allows one to increase the intrinsic breakdown voltage of the drain junction BV_{CBO}, while the LDS (lightly doped source) structure reduces the emitter efficiency of the parasitic bipolar transistor [8.13-8.17]. In addition, deep submicron devices (*e.g.* L=0.25 μm) will benefit from the reduction of supply voltage to 2...3.3 volts. At these voltages, drain impact ionization is significantly reduced, and parasitic bipolar effects tend to disappear. The second point of concern for volume production of SOI integrated circuits on a commercial basis is the defect density in SOI films, which is larger than in bulk silicon. High-quality bulk silicon wafers present defect densities on the order of 10 dislocations/cm^2. A quick look at Figure 2.7.6 shows that state-of-the-art SIMOX wafers have 10^3 cm^{-2} dislocation density, but that the situation is rapidly evolving and that defect densities similar to bulk can be expected in a near future. It is also worthwhile giving some thoughts to the effects produced by defects in MOS devices. These can be of different natures, such as degradation of the gate oxide quality or junction leakage. In a bulk device, a defect can produce a yield hazard if it is located in the active area, *i.e.* if it is located in the channel region, the source or the drain. If λ is the minimum design rule for all levels, the area where the presence of a defect can be felt is given by: A = device length x device width = [2(field isolation to contact hole spacing + contact hole length + contact hole to gate spacing) + gate length)] x [field isolation to contact hole spacing + contact hole width + field isolation to contact hole spacing] = $[2(3\lambda)+\lambda][3\lambda]=21\lambda^2$. In the case of a thin-film SOI device with reach-through junctions, the area where the presence of a defect can be felt is given by the area of the channel region only, because of the full dielectric isolation of the source and drain junctions. This area is equal to $3\lambda^2$ in our example, and is, therefore, significantly smaller than in a bulk device. According to such considerations, the use of SIMOX wafers permits one to obtain higher fabrication yield than bulk silicon if the defect densities in both materials are comparable.

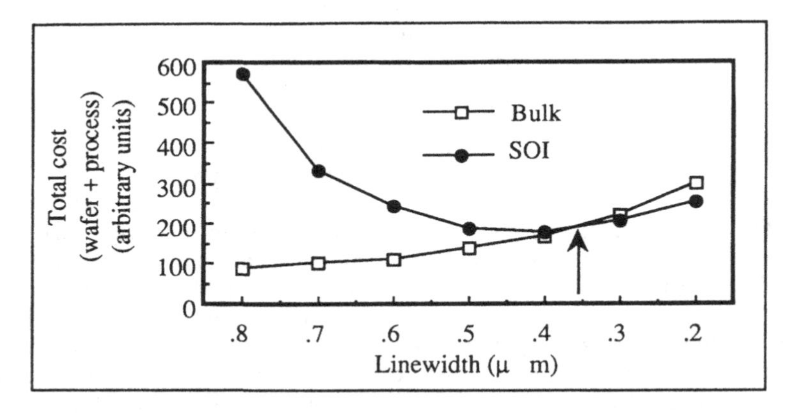

Figure 8.2.1: Predicted cost evolution of bulk and SIMOX CMOS wafers (wafer cost + CMOS processing cost) as a function of minimum device dimension. The arrow indicates the break-even point where SOI technology becomes economically favorable.

Figure 8.2.1 presents a generally accepted scenario for the cost evolution of fully processed bulk and thin-film SOI CMOS wafers. The cost of a virgin bulk wafer is low and will remain so in the future (though it might increase if epi wafers are needed for deep-submicron devices). The cost of SIMOX wafers is much larger (up to 40 times higher), but the increased demand stimulates concurrence between suppliers and the increased throughput of the upcoming generation of oxygen implanters (300 mA-class) will force the wafer price to drop to 3-5 times the price of a bulk wafer. The cost of front-end CMOS processing, on the other hand, increases steadily as smaller and smaller devices are fabricated. This cost increase is significantly larger in bulk CMOS processing than in thin-film SOI processing. Indeed, deep-submicron CMOS processing necessitates heavy investments in terms of R&D efforts in order to produce adequate device isolation, shallow junctions, anti-latchup and anti-punchthrough techniques. SOI devices, on the other hand, are scalable in their current form to deep-submicron dimensions without the need for any major new technique development, which reduces the need for costly R&D investments. Front-end thin-film SOI processing is already currently cheaper than front-end bulk processing, and it will become even more so in the future. When the cost of both substrate and processing are added, curves such as represented in Figure 8.2.1 are obtained, in which one observes that thin-film SOI technology becomes less expensive than classical bulk technology in the deep-submicron regime. The break-even point appears to be located around 0.3 μm device dimensions, and is indicated by an arrow in Figure 8.2.1 [8.18].

Application	Ref.
Large CMOS circuits	
16k SRAM	8.19
64k SRAM	8. 20
256k SRAM	8.21
16-bit processor	8.22
High-speed CMOS circuits	
21 ps/gate ring oscillator	8.23
2 GHz frequency divider	8.24
1.5 GHz ECL-compatible output buffer	8.25
6.2 GHz prescaler	8.26

Table 8.2.1: Some VLSI and high-speed SOI circuits

Large circuits, such as 16k, 64k and 256k SRAMs have been realized on SOI substrates (Table 8.2.1). Some of these chips are commercially available. The realization of these VLSI circuits demonstrates the high quality of today's SOI materials. Furthermore, the fabrication yield of these SRAMs is sometimes better than that of the corresponding bulk parts.

Another definite advantage of SOI technology is the increase of circuit speed due to the reduction of source and drain capacitances and, in the case of fully-depleted SOI devices, to the increase of transconductance. In 1988 and 1989, small thin-film SOI circuits (frequency dividers and prescalers) broke the CMOS world speed record (Table 8.2.1). Figure 8.2.2 compares the propagation delay in some state-of-the art CMOS ring oscillators from the literature as a function of the gate length x gate oxide thickness product [8.18]. The SOI oscillators show a steady 50% improvement of gate delay over the bulk devices. Larger circuits, such as SRAMs and microprocessors, show 15% to 20% improvement of access time or maximum operating frequency, even if the design is directly transferred from bulk to SOI without optimization [8.27].

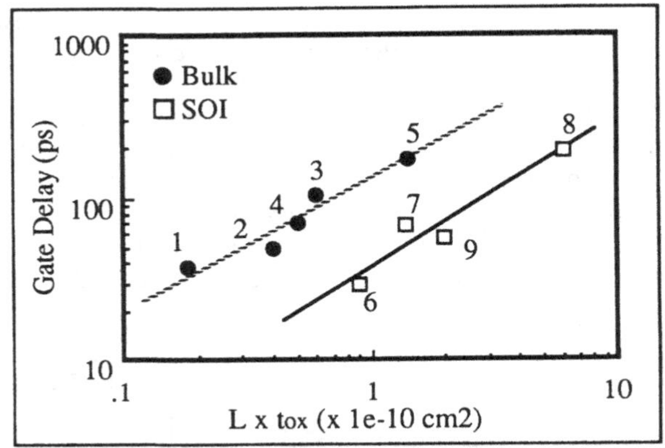

Figure 8.2.2: Delay per gate in advanced bulk and SOI CMOS ring oscillators as a function of the product of gate length and gate oxide thickness. The references are: 1-IBM, 1988; 2-AT&T, 1988, 3: TI, 1988; 4-AT&T, 1989; 5-TI, 1988, 6-Hughes, 1988; 7-HP, 1988; 8-Mitsubishi, 1988; 9-Northern Telecom 1989.

In addition to the increase in speed, SOI technology permits one to reduce both the standby power consumption and the operating power consumption of the circuits. The standby power consumption is lower than in bulk devices because of the reduced area of the reach-through SOI source and drain junctions. Standby currents of 10 nA and even 6.5 nA have been obtained in 64k and 256k SRAMs, respectively (V_{DD}=5V) [8.20, 8.21]. The operating power consumption is reduced mainly by the lower value of the source and drain capacitances. Indeed, a 50% reduction of the current consumption is observed in SOI 16k SRAMs, compared to the equivalent bulk parts (for a constant access time) [8.28]. Similarly, a 30% decrease of access time is obtained if the current consumption is maintained constant in the comparison between bulk and SOI devices [8.28]. From another standpoint, it can be said that a generation of CMOS SOI delivers the same performances as the next generation of bulk circuits [8.18].

8.3. Three-dimensional integrated circuits

The circuits described so far are classical two-dimensional arrangements of transistors, exactly like in bulk technology. SOI transistors can, however, be used as elementary building blocks to fabricate three-dimensional integrated circuits. 3-D circuits are based on the stacking of several active device layers upon one another. The first (bottom) layer is usually realized in a bulk silicon wafer using classical bulk device processing techniques. An insulating layer of silicon dioxide is then deposited, followed by a layer of polysilicon. Laser recrystallization or eventually e-beam recrystallization is then used to melt the polysilicon layer and recrystallize it into device-worthy silicon. Laser and e-beam recrystallization are used preferentially to other SOI techniques such as ZMR because the laser beam heats only the superficial polysilicon layer, and because the dwell time (milliseconds) is short enough to avoid degradation of the devices already fabricated below the layer being recrystallized. LSPE and ELO are also techniques which, in the long run, present a potential for the fabrication of 3-D integrated circuits. The first steps towards 3D integration were based on the realization of 3D devices, such as compact CMOS logic gates [8.29-8.31]. Figure 8.3.1 presents the example of a stacked CMOS inverter where the n-channel device is realized in a bulk silicon substrate. After fabrication of this n-channel MOSFET, a gate oxide is grown on the polysilicon gate, and polysilicon is deposited, recrystallized using a laser, and doped to form a p-channel transistor. Both transistors share a joint gate, and their drains are connected together in order to form a CMOS inverter. Epitaxial lateral overgrowth (ELO) can be used as an alternative to laser recrystallization to fabricate such a structure [8.31].

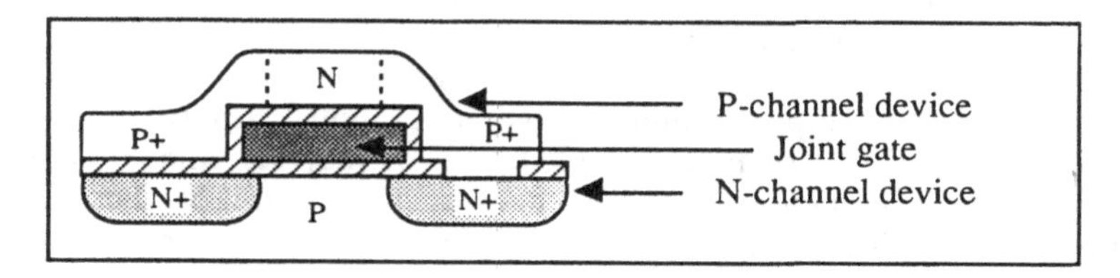

Figure 8.3.1: Stacked CMOS inverter.

Stacked CMOS structures achieve an increase of packing density compared to standard CMOS, but they do not lend themselves to the fabrication of more than two active layers. Much higher flexibility and modularity are obtained by stacking active layers which are electrically independent from one another. In this approach, which is currently used for all 3-D circuits, the different active layers are insulated from one another by a thick (≅ 1 μm) insulating layer (usually SiO_2). Sometimes, the interlevel insulator is composed of a SiO_2/polysilicon

$/SiO_2$ sandwich in which the polysilicon serves two purposes: firstly, its presence reduces the crosstalk between different active layers (it provides an interlevel ground plane) and, secondly, it facilitates the laser recrystallization step by improving the uniformity of the reflectivity of the multilayer structure across the wafer. An example of cross-section through a 3-D integrated circuit is shown in Figure 8.3.2. This example outlines the different challenges faced by the fabrication of 3-D integrated circuits. These are: the formation of stacked device-quality silicon layers, the planarization of the dielectric insulator between the active layers, the development of a low-resistivity, temperature-resistant interlayer interconnection system, and the development of 3-D design tools and design methodologies.

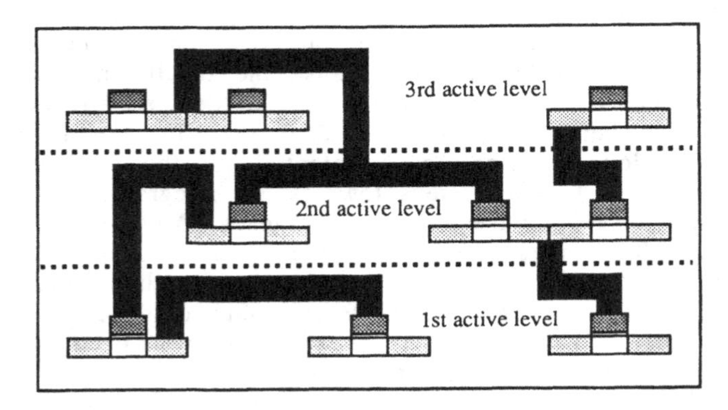

Figure 8.3.2: Cross-section of a 3-D integrated circuit.

Three-dimensional integrated circuit technology is particularly well suited for the realization of large systems involving massive parallel processing. To illustrate this, let us consider te example of the image processing chip (computer vision system) described in Figure 8.3.3. This circuit extracts the boundaries of objects and transfers this information to a computer which recognizes the nature of the object and its orientation. The circuit operates in the following way. Image detection is performed by a bidimensional photodiode array. The information contained in each pixel is digitized (we will consider only two luminosity levels, black and white, *i.e.* "1" and "0") and each pixel is compared with its eight neighbors (north, north-east, east, south-east, south, south-west, west, and north-west) to find out whether the point it represents lies inside the object, outside of it, or on its boundary. Such a comparison can be realized by a simple software routine or hard-wired by means of logical XOR gates. When the edges of the object have been extracted, they can be expressed in the form of 2D vectors from which the nature and the in-plane orientation of the object can be deduced. In the case of the implementation of such a system using classical 2-D chips, the data contained in each pixel must be transferred after digitization to a computer which performs the comparison operations (on-chip digitization is possible, but at the expense of the sensor fill factor). Because of the limited amount of I/O pins of the circuits, this data transfer is carried out in a largely serial manner. If the amount of data is large (1024 x 1024 pixel array = 1 Mbit of data for each frame), the serial data transfer process quickly becomes the bottleneck of the system, due to the limited bandwidth of the data transmission bus [8.32]. Once in the computer, the image is reconstructed in the memory the comparison and feature extraction operations are performed.

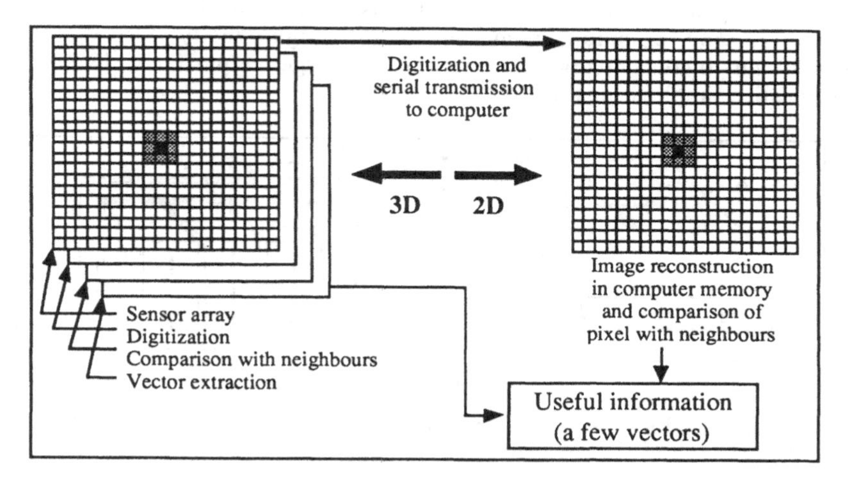

Figure 8.3.3: comparison between 3-D parallel data processing and 2-D serial data processing.

In the case of a 3-D implementation of such a system, sensing, digitization, comparison and vector extraction can be carried out in four separate active device layers (Figure 8.3.3). All the pixels can be transferred *simultaneously* and *vertically* from the sensor array to the digitization layer. In this way, maximum sensor fill factor is obtained, which minimizes the integration time of the photodiode current. After A to D conversion, the data are transferred *simultaneously* and *vertically* to the next layer where comparison operations between pixels are performed. It is worth noting that each pixel is naturally surrounded by the pixels with which it has to be compared. Once this comparison operation has been performed, the data can be transferred to the bottom layer for feature extraction. The important advantage of the 3-D approach is the highly parallel inter-layer data transfer, which eliminates the data transmission bottleneck experienced in the 2D implementation. As a result, considerably higher speed performance is obtained from 3-D parallel processing systems. An increase of real-time image processing by a factor of about 1000 (compared to conventional 2D processing) has been observed in small (5x5 pixels) systems [8.33], and such processing speed advantage is only expected to increase as the complexity of the system increases [8.34]. It is also worthwhile noting that the length of the interconnections between two different active layers is typically on the order of a micrometer, while the length of interconnections between different logic blocs in 2-D circuits is generally several hundred micrometers. This also gives rise to further processing speed increase in 3-D systems. Massively parallel processing capability and reduced interconnection length offer advantage not only in image processing circuits, but also in neural network [8.35], associative and common memories [8.36], high-density SRAMs [8.37], three-dimensional sensing of impinging energetic particles [8.38], and master slice applications [8.39].

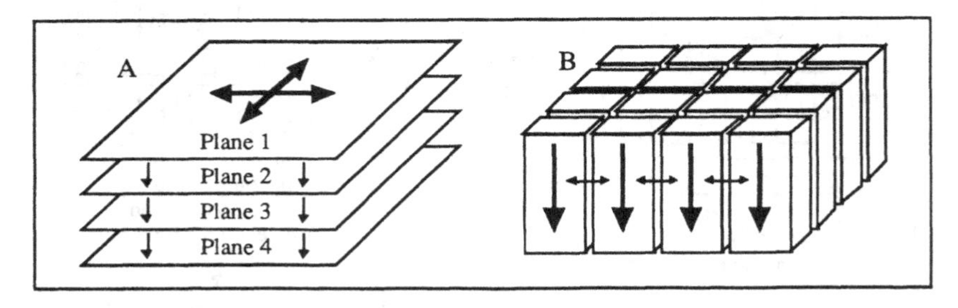

Figure 8.3.4: 3-D arrangement of 2-D circuits (A) and 2-D arrangement of 3-D circuits (B).

Owing to the nature of the technological steps using which it is fabricated, a three-dimensional IC can be viewed as a succession of 2-D circuits stacked one upon another. Functionally, it can be considered either as a 3-D association of 2-D processing planes or a 2-D association of vertical data flow paths (Figure 8.3.4) [8.34]. In the first case, most of the data processing takes place within an horizontal plane and is then transferred to the next layer (3-D master slice chips belong to this category). In the case of the 2-D association of vertical circuits, the data flow from top to bottom in a highly parallel fashion, but interaction such as the comparison of a pixel with its neighbors between data occurs within horizontal planes. Most image processing 3-D integrated circuits belong to this category. One can, of course also conceive fully 3-D circuits where there is no preferential (neither horizontal nor vertical) direction for the data flow (i.e. "brain-like" neural networks), but the realization of such systems will necessitate revolutionary design tools and methodologies which are yet to be invented.

We will now illustrate the advantages of 3-D integrated circuits by two examples: a moving object detector [8.40] and a character recognition system [8.41, 8.42]. The moving object detector consists of the following layers (from top to bottom): an optical sensor array consisting of 64 x 64 amorphous silicon photodiodes (fourth floor), a 64 x 64 level-detector array which digitizes each pixel (third floor), a first 64 x 64 bit frame memory (FM1 - second floor), and finally a last layer consisting of a second 64 x 64 bit frame memory (FM2) and subtracting logic used to compare the information contained in FM1 and FM2 (first floor) [8.40]. The device operates as follows (Figure 8.3.5). In a first step, an optical image is focused on the amorphous silicon sensor matrix in the top layer. After suitable integration time, the sensed image is digitized into a one-bit code per pixel in the third floor. The threshold between "black" and "white" can be adjusted by means of an external reference voltage. Each digitized pixel is then written in FM2. In a second step, the next image is sensed by the top layer, digitized in the third floor, and written in FM1. In a third step, all the pixels of the first frame memory are simultaneously subtracted from the corresponding pixels of second frame memory, and an output signal is sent. If the object sensed by the circuit has not moved, the contents of FM1 and FM2 will be identical, and the output signal will be zero. If there has been movement, a differential non-zero signal (positive local-time derivative) is produced.

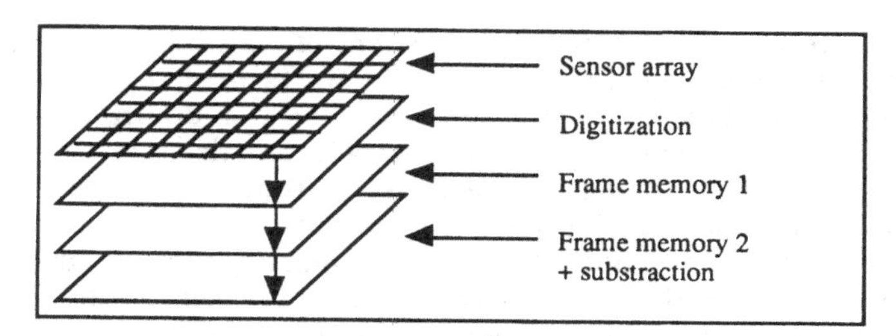

Figure 8.3.5: 3-D moving object sensor.

Another example illustrating the possibilities offered by 3-D integration technology can be found in the field of character recognition systems. A first implementation of such a device was first realized in 1988 [8.41], and an improved version was reported in 1990 [8.42]. In its first version, the circuit consists of three active layers (Figure 8.3.6): a 210-pixel SOI photodiode array, a "converter block" consisting of a digitizer and a data hold block, and a comparator block consisting of a ROM containing 64 alphanumeric characters and the logic necessary for comparing the hold data with the characters stored in the ROM. The circuit operates as follows: the image of a character written on paper is focussed on the photodiode array, digitized and held in the second layer.

The held data is automatically rewritten if the incoming image changes during the system operation. The content of the hold data memory is then transferred, row by row, to the bottom layer, where comparison between the sensed characters and those stored in the ROM can be carried out. Correction logic is included in the comparator block, such that imperfect characters, such as the "S" of Figure 8.3.6 can be recognized. If a match is found, the ASCII code corresponding to the recognized letter is produced at the output of the system. Owing to its highly parallel processing mechanism this circuit performs character recognition approximately ten times faster than an equivalent 2-D system using serial processing [8.41].

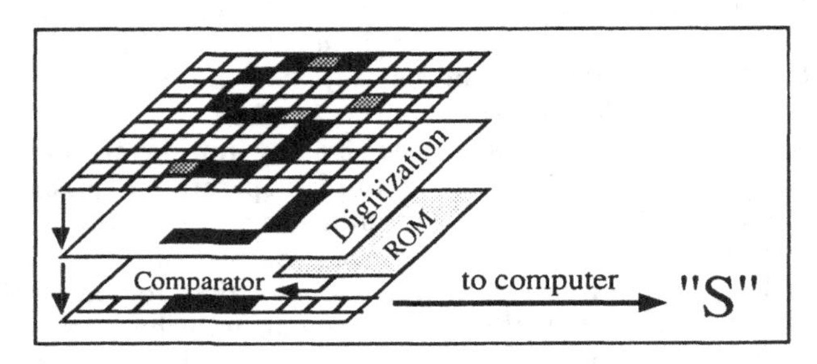

Figure 8.3.6: 3-D Character recognition chip.

More recently, an improved version of this character recognition chip has been reported [8.42]. The circuit is realized in four active silicon layers: a photodiode array (fourth floor), a digitizer (third floor), a queue register (second floor), and a data comparison layer (first floor). The data are transferred in an asynchronous manner from the photodiodes through the digitizer to the queue memory. If the sensed character is not well positioned for comparison with the reference characters stored in the associative memory located on the first floor, the data held in the second floor are automatically shifted in one of eight possible directions (north, northeast, east, ...), and the missing bits resulting from the shift operation are masked (*e.g.* the left column is masked if the image has been shifted to the right). Masked bits are not taken into account during the subsequent comparison with the characters stored in the associative memory. In addition to the shifting operation, a second technique for improving the character recognition mechanism has been built in the system: if an imperfect character is detected, the defective part of the character is masked. The masking command is generated by a command from the mask register on the second floor. Only the "perfect" portion of the sensed character is the compared with the characters stored in the associative memory. This technique improves the character recognition yield of the system.

BIBLIOGRAPHY

Chapter 1

1.1. J.E. Lilienfield, U.S. patents 1,745,175 (filed 1926, issued 1930), 1,877,140 (filed 1928, issued 1932), and 1,900,018 (filed 1928, issued 1933)

1.2. see for example: W. Shockley, "The path to the conception of the junction transistor", IEEE Trans. on Electron Devices, Vol. ED-23, no. 7, p. 597, July 1976

1.3. see for example: D. Kahng, "A historical perspective on the development of MOS transistors and related devices", IEEE Trans. on Electron Devices, Vol. ED-23, no. 7, p. 655, July 1976

1.4. R.R. Troutman, Latchup in CMOS Technology, Kluwer Academic Publishers, 1986

1.5. T.W. MacElwee, Ext. Abstracts of the Electrochem. Society Spring Meeting, Los Angeles, p. 280, 1989

Chapter 2

2.1. I. Golecki, in "Comparison of thin-film transistor and SOI technologies", Ed. by H.W. Lam and M.J. Thompson, Mat. Res. Soc. Symp. Proc., Vol. 33, p. 3, 1984

2.2. T. Sato, J. Iwamura, H. Tango, and K. Doi, in "Comparison of thin-film transistor and SOI technologies", Ed. by H.W. Lam and M.J. Thompson, Mat. Res. Soc. Symp. Proc., Vol. 33, p. 25, 1984

2.3. H.M. Manasevit and W.I. Simpson, J. Appl. Phys, Vol.35, p. 1349, 1964

2.4. RCA part number MWS-5101

2.5. B.E. Forbes, Hewlett-Packard Journal, p. 2, 1977

2.6. R.J. Hollingsworth, A.C. Ipri, and C.S. Kim, IEEE J. Solid-State Circuits, Vol. SC-13, p. 664, 1978

2.7. A.G.F. Dingwall, R.G. Stewart, B.C. Leung, and R.E. Stricker, Proc. IEDM, p. 193, 1978

2.8. D. Adams, D. Uehara, D. Wheeler, and D. Williams, Proc. of the IEEE SOS/SOI Technology Workshop, p. 58, 1987

2.9. A.E. Schmitz, R.H. Walden, M. Montes, D. M. Courtney, and E. Stevens, Tech. Digest of Symposium on VLSI Technology, p. 67, 1988

2.10. D.J. Dumin, S. Dabral, M.H. Freytag, P.J. Robertson, G.P. Carver, and D.B. Novoty, IEEE Trans. on Electron. Dev., Vol. 36, p. 596, 1989

2.11. A.C. Ipri, Applied Solid-State Sciences, Supplement 2, Silicon Integrated Circuits, Part A, Ed. by. D. Kahng, Academic Press, p. 253, 1981

2.12. M.S. Abrahams and C.J. Buiocchi, Appl. Phys. Lett., Vol. 20, p. 91, 1972

2.13. H.M. Manasevit, I. Golecki, L.A. Moudi, J.J. Yang, and J.E. Mee, J. Electrochem. Soc., Vol. 130, p. 1752, 1983

2.14. G.D. Robertson, Jr., P.K. Vasudev, R.G. Wilson, and V.R. Deline, Appl. Surf. Sci., Vol. 14, p. 128, 1982/83

2.15. G. Yaron and L.D. Hess, Solid-State Electronics, Vol. 23, IEEE Trans. on Electron Devices, Vol. 27, p. 573, 1980

2.16. S.S. Lau, S. Matteson, J.W. Mayer, P. Revesz, J. Gyulai, J.Roth, T.W. Sigmond, and T. Cass, Appl. Phys. Lett, Vol. 34, p. 76, 1979

2.17. J. Amano and K. Carey, Appl. Phys. Lett., Vol. 39, July 1980

2.18. P.K. Vasudev and D.C. Mayer, in "Comparison of thin-film transistor and SOI technologies", Ed. by H.W. Lam and M.J. Thompson, Mat. Res. Soc. Symp. Proc., Vol. 33, p. 35, 1984

2.19. M.E. Roulet, P. Schwob, I. Golecki, and M.A. Nicolet, Electronics Letters, Vol. 15, p. 527, 1979

2.20. I. Golecki, H.M. Manasevit, L.A. Moudy, J.J. Yang, and J.E. Mee, Appl. Phys. Letters, Vol. 42, p. 501, 1983

2.21. I. Golecki, R.L. Maddox, H.L. Glass, A.L. Lin, and H.M. Manasevit, presented at the 26th electronic Materials Conf., Santa Barbara, CA, June 1984

2.22. M. Ihara, Y.Arimoto, M. Jifiku, T. Kimura, S. Kodama, H. Yamawaki, and T. Yamaoka, J. Electrochem. Soc., Vol. 129, p. 2569, 1982

2.23. Y. Hokari, M. Mikami, K. Egami, H. Tsuya, and M. Kanamori, Techn. Digest of IEDM, p. 368, 1983

2.24. K. Ikeda, H. Yamawaki, T. Kimura, M. Ihara, and M. Ozeki, Ext. Abstr. 5th International Workshop on Future Electron Devices, Miyagi-Zao, Japan, p. 225, 1988

2.25. T. Asano and H. Ishiwara, Jpn. J. Appl. Phys., Vol. 21, suppl. 21-1, p. 187, 1982

2.26. T.R. Harrison, P.M. Mankiewich, and Dayem, Appl. Phys. Lett., Vol. 41, p. 1102, 1982

2.27. H. Onoda, M. Sasaki, T. Katoh, and N. Hirashita, IEEE Trans. Electron Dev., Vol. ED-34, p. 2280, 1987

2.28. S. Sugiura, T. Yoshida, and K. Shono, Jpn. J. Appl. Phys, Vol. 22, p. L426, 1983

2.29. T. Takenaka and K. Shono, Jpn. J. Appl. Phys, Vol. 13, p. 1211, 1974

2.30. M. Morita, S. Isogai, N. Shimizu, K. Tsubouchi, and N. Mikoshiba, Jpn. J. Appl. Phys, Vol. 20, p. L173, 1981

2.31. D.M. Jackson, Jr. and R.W. Howard, Trans. Met. Soc. AIME, Vol. 223, p. 468, 1965

2.32. H.M. Manasevit, D.H. Forbes, and I.B. Cardoff, Trans Met. Soc. AIME, Vol. 236, p. 275, 1966

2.33. G. Shimaoka and S.C. Chang, J. Vac. Sci. Technol., Vol. 9, p. 235, 1972

2.34. C.H. Fa and T.T. Jew, IEEE Trans. on Electron Devices, Vol. ED-13, p. 290, 1966

2.35. T.I.Kamins, Solid-State Electronics, Vol. 15, p. 789, 1972

2.36. S.W. Depp, B.G. Huth, A. Juliana, and R.W. Koepcke, in "Grain Boundaries in Semiconductors",Ed. by H.J. Leamy, G.E. Pike, and C.H. Seager, Mat. Res. Soc. Symp. Proceedings, Vol. 5, p. 297, 1982

2.37. T.I. Kamins and P.J. Marcoux, IEEE Electron Device Letters, Vol. EDL-1, p. 159, 1980

2.38. H. Shichiro, S.D.S. Malhi, P.K. Chatterjee, A.H. Shah, G.P. Pollack, W.H. Richardson, R.R. Shah, M.A. Douglas, and H.W. Lam, in "Comparison of thin-film transistor and SOI technologies", Ed. by H.W. Lam and M.J. Thompson, Mat. Res. Soc. Symp. Proceedings, North-Holland, Vol. 33, p. 193, 1984

2.39. S.D.S. Malhi, in "Comparison of thin-film transistor and SOI technologies", Ed. by H.W. Lam and M.J. Thompson, Mat. Res. Soc. Symp. Proceedings, North-Holland, Vol. 33, p. 147, 1984

2.40. G.K Celler, H.J. Leamy, L.E. Trimble, and T.T. Sheng, Appl. Phys Lett., Vol. 39, p. 425, 1981

2.41. C. Hill, in"Laser and Electron-Beam Solid Interactions and Material Processing", Ed. by J.F. Gibbons, L.D. Hess, and T.W. Sigmon, Mat. Res. Soc. Symp. Proceedings, Vol. 1, p. 361, 1981

2.42. N.M. Johnson, D.K, Biegelsen, H.C. Tuan, M.D. Moyer, and E. Fennel, IEEE Electron Dev. Lett, Vol. EDL-3, p. 369, 1982

2.43. G.J. Willems, J.J. Poortmans, and H.E. Maes, J. Appl. Phys., Vol. 62, p. 3408, 1987

2.44. T.I. Kamins, M.M. Mandurah, and K.C. Saraswat, J. Electrochem. Soc., Vol. 125, p. 927, 1978

2.45. H.W. Lam, R.F. Pinizotto and A.F. Tasch, J. Electrochem. Soc., vol 128, p. 1981, 1981

2.46. K.F. Lee, T.J. Stultz, and J.F. Gibbons, Semiconductors and Semimetals, Vol. 17, cw Processing of Silicon and other Semiconductors, Academic Press, p.227, 1984

2.47. M. Ohkura, K. Kusukawa, I. Yoshida, M. Miyao, and T. Tokuyama, Ext. Abstr. of the 15th Conf. on Solid-State Devices and Materials, Japan Society of Appl. Physics, p. 43, 1983

2.48. H.P. Le and H.W. Lam, Ext. Abstr. of Electrochem. Soc. Spring Meeting, Vol. 82-1, p. 240, 1982

2.49 D.J. Wouters and H.E. Maes, J. Appl. Phys, Vol. 66, p. 900, 1989

2.50. J.P. Colinge, H.K. Hu, and S. Peng, Electronics Letters, Vol. 21, p. 1102, 1985

2.51. P.Seegebrecht, Ext. Abstracts of 5th Internat. Workshop on Future Electron Devices, Miyagi-Zao, Japan, p. 19, 1988

2.52. J.M. Hodé, J.P. Joly, and P. Jeuch, Ext. Abstr. of Electrochem. Soc. Spring Meeting, Vol. 82-1, p. 232, 1982

2.53. S. Kawamura, J. Sakurai, M. Nakano, and M. Tagaki, Appl. Phys. Lett., Vol. 40, p. 232, 1982

2.54. N.A. Aizaki, Appl. Phys. Lett., Vol. 44, p. 686, 1984

2.55. J.P. Colinge, E. Demoulin, D. Bensahel, and G. Auvert, Appl. Phys. Lett., Vol. 41, p. 346, 1982

2.56. J.P. Colinge, Ext. Abstracts of 2nd Internat. Workshop on Future Electron Devices, Shuzenji, Japan, p. 13, 1985

2.57. T. Morishita, T. Miyajima, J. Kudo, M. Koba, and K. Awane, Ext. Abstracts of 2nd Internat. Workshop on Future Electron Devices, Shuzenji, Japan, p. 35, 1985

2.58. E. Fujii, K. Senda, F. Emoto, and Y. Hiroshima, Appl. Phys. Lett., Vol. 63, p. 2633, 1988

2.59. K. Sugahara, S. Kusunoki, Y. Inoue, T. Nishimura, and Y. Akasaka, J. Appl. Phys., Vol. 62, p. 4178, 1987

2.60. T. Nishimura, Y. Inoue, K. Sugahara, S. Kusunoki, T. Kumamoto, S. Nakagawa, M. Nakaya, Y. Horiba, and Y. Akasaka, Proceedings of International Electron Device Meeting, p. 111, 1987

2.61. T. Kunio, K. Omaya, Y. Hayashi, and M. Morimoto, Proceedings of International Electron Device Meeting, p. 837, 1989

2.62. Y. Akasaka, Ext. Abstracts of 8th Internat. Workshop on Future Electron Devices, Kochi, Japan, p. 9, 1990

2.63. D. Bensahel, Ext. Abstracts of 5th Internat. Workshop on Future Electron Devices, Miyagi-Zao, Japan, p. 9, 1988

2.64. D.K. Biegelsen, N.M. Johnson, D.J. Bartelink, and M.D. Moyer, Appl. Phys. Lett., Vol. 38, p. 150, 1981

2.65. S. Akiyama, N. Noshii, K. Yamazaki, M. Yoneda, S. Ogawa, and Y. Terui, Ext. Abstracts of 2nd Internat. Workshop on Future Electron Devices, Shuzenji, Japan, p. 41, 1985

2.66. R. Mukai, N. Sasaki, T. Iwai, S. Kawamura, and M. Nakano, Proceedings of International Electron Device Meeting, p. 360, 1983

2.67. A.J. Auberton-Hervé, J.P. Joly, P. Jeuch, J. Gautier, and J.M. Hodé, Proceedings of International Electron Device Meeting, p. 808, 1984

2.68. S. A. Lyon, R. J. Namanich, N.M. Johnson, and D.K. Biegelsen, Appl. Phys. Lett., Vol. 40, p. 316, 1982

2.69. Y. Itoh, A. Wada, K. Morimoto, and K. Yamazaki, Ext. Abstracts of 8th Internat. Workshop on Future Electron Devices, Kochi, Japan, p. 97, 1990

2.70. D.K. Biegelsen, Ext. Abstracts of 2nd Internat. Workshop on Future Electron Devices, Shuzenji, Japan, p. 31, 1985

2.71. K. Egami and M. Kimura, Ext. Abstracts of 2nd Internat. Workshop on Future Electron Devices, Shuzenji, Japan, p. 105, 1985

2.72. R.C. McMahon, Microelectronic Engineering, Vol. 8, p. 255, 1988

2.73. T. Hamasaki, T. Inoue, I. Higashinakagawa, T. Yoshii, and H. Tango, J. Appl. Phys., Vol. 59, p. 2971, 1986

2.74. K. Shibata, T. Inoue, T. Takigawa, and S. Yoshii, Appl. Phys. Lett., Vol. 39, p. 645, 1981

2.75. J.A. Knapp and S.T. Picraux, J. Crystal Growth, Vol. 63, p. 445, 1983

2.76. T. Yoshii, T. Hamasaki, K. Suguro, T. Inoue, M. Yoshimi, K. Taniguchi, M. Kashiwagi, and H. Tango, Ext. Abstracts of 2nd Internat. Workshop on Future Electron Devices, Shuzenji, Japan, p. 51, 1985

2.77. M. Yoshimi, H. Hazama, M. Takahashi, S. Kambayashi, M. Kemmochi, and H. Tango, Ext. Abstracts of 5th Internat. Workshop on Future Electron Devices, Miyagi-Zao, Japan, p. 143, 1988

2.78. J.R. Davis, R.A. McMahon, and H. Ahmed, Electronics Letters, Vol. Vol. 18, p. 163, 1982

2.79. J.C.C. Fan, M.W. Geis, and B.Y. Tsaur, Appl. Phys. Lett., Vol. 38, p. 365, 1981

2.80. M.W. Geis, H.I. Smith, B.Y. Tsaur, J.C.C. Fan, D.J. Silversmith, R.W. Mountain, and R.L. Chapman, in "Laser-Solid Interactions and Transient Thermal Processing of Materials", Narayan, Brown and Lemons Eds., (North-Holland), MRS Symposium Proceedings, Vol. 13, p. 477, 1983

2.81. A. Kamgar and E. Labate, Mat. Letters, Vol. 1, p. 91, 1982

2.82. T.J Stultz, Appl. Phys. Lett., Vol. 41, p. 824, 1982, and
T.J. Stultz, J.C. Sturm, and J.F. Gibbons, in "Laser-Solid Interactions and Transient Thermal Processing of Materials", Narayan, Brown and Lemons Eds., (North-Holland), MRS Symposium Proceedings, Vol. 13, p. 463, 1983

2.83. D.P. Vu, M. Haond, D. Bensahel, and M. Dupuy, J. Appl. Phys., Vol. 54, p. 437, 1983

2.84. P.W. Mertens, D.J. Wouters, H.E. Maes, A. De Veirman, and J. Van Landuyt, J. Appl. Phys., Vol. 63, p. 2660, 1988

2.85. D. Dutartre, in "Silicon-On-Insulator and Buried Metals in Semiconductors", Sturm, Chen, Pfeiffer and Hemment Eds., (North-Holland), MRS Symposium Proceedings, Vol. 107, p. 157, 1988, and
D. Dutartre, M. Haond, and D. Bensahel, in "Semiconductor-On-Insulator and Thin Film Transistor Technology", Chiang, Geis and Pfeiffer Eds., (North-Holland), MRS Symposium Proceedings, Vol. 53, p. 89, 1986

2.86. D. Dutartre, Appl. Phys. Lett., Vol. 48, p. 350, 1986

2.87. H.A. Atwater, Jr., VLSI Memo 83-149, Dept. of Electr. Eng. and Computer Science, MIT, Cambridge, MA, p. 23, 1983

2.88. H.J. Leamy, C.C. Chang, H. Baumgard, R.A. Lemons, and J. Cheng, Mat. Lett., Vol. 1, p. 33, 1982

2.89. R.A. Lemons, M.A. Bosch, and D. Herbst, in "Laser-Solid Interactions and Transient Thermal Processing of Materials", Narayan, Brown and Lemons Eds., (North-Holland), MRS Symposium Proceedings, Vol. 13, p. 581, 1983

2.90. M.W. Geis, H.I. Smith, B.Y. Tsaur, J.C.C. Fan, D.J. Silversmith, and R.W. Mountain, J. Electrochem. Soc., Vol. 129, p. 2812, 1982

2.91. H.F. Wolf, "Semiconductor Data", Pergamon Press, Oxford, 1969

2.92. J.C.C. Fan, B.Y Tsaur ,C.K. Chen, J.R. Dick, and L.L. Kazmerski, in "Energy Beam-Solid Interactions and Transient Thermal Processing", Fan and Johnson Eds., (North-Holland), MRS Symposium Proceedings, Vol. 23, p. 477, 1984

2.93. M.W. Geis, H.I. Smith, D.J. Silversmith, R.W. Mountain, and C.V. Thomson, J. Electrochem. Soc., Vol. 130, p. 1178, 1983

2.94. L. Pfeiffer, S. Paine, G.H. Gilmer, W. Saarloos, and K.W. West, Phys. Rev. Lett., Vol. 54, p. 1944, 1985

2.95. L. Pfeiffer, K.W. West, D.C. Joy, J.M. Gibson, and A.E. Gelman, in "Semiconductor-On-Insulator and Thin Film Transistor Technology", Chiang, Geis and Pfeiffer Eds., (North-Holland), MRS Symposium Proceedings, Vol. 53, p. 29, 1986

2.96. J.S. Im, C.K. Chen, C.V. Thompson, M.W. Geis, and H. Tomita, in "Silicon-On-Insulator and Buried Metals in Semiconductors", Sturm, Chen, Pfeiffer and Hemment Eds., (North-Holland), MRS Symposium Proceedings, Vol. 107, p. 169, 1988

2.97. P.M. Zavaracky, D.P. Vu, L. Allen, W. Henderson, H. Guckel, J.J. Sniegowski, T.P. Ford, and J.C.C. Fan, in "Silicon-On-Insulator and Buried Metals in Semiconductors", Sturm, Chen, Pfeiffer and Hemment Eds., (North-Holland), MRS Symposium Proceedings, Vol. 107, p. 213, 1988

2.98. M.W. Geis, C.K. Chen, H.T. Smith, P.M. Nitishin, B.Y. Tsaur, and R.W. Mountain, in "Semiconductor-On-Insulator and Thin Film Transistor Technology", Chiang, Geis and Pfeiffer Eds., (North-Holland), MRS Symposium Proceedings, Vol. 53, p. 39, 1986

2.99. E.W. Maby, M.W. Geis, Y.L. LeCoz, D.J. Silversmith, R.W. Mountain, and D.A. Antoniadis, Electron Dev. Lett., Vol. 2, p. 241, 1981

2.100. C.K. Chen, L. Pfeiffer, K.W. West, M.W. Geis, S. Darack, G. Achaibar, R.W. Mountain, and B.Y. Tsaur, in "Semiconductor-On-Insulator and Thin Film Transistor Technology", Chiang, Geis and Pfeiffer Eds., (North-Holland), MRS Symposium Proceedings, Vol. 53, p. 53, 1986

2.101. M. Haond, D. Dutartre, R. Pantel, A. Straboni, and B. Vuillermoz, in "Semiconductor-On-Insulator and Thin Film Transistor Technology", Chiang, Geis and Pfeiffer Eds., (North-Holland), MRS Symposium Proceedings, Vol. 53, p. 59, 1986

2.102. E. Yablonovitch and T. Gmitter, J. Electrochem. Soc., Vol. 131, p. 2625, 1984

2.103. P.W. Mertens and H.E. Maes, in "Beam-Solid Interactions: Physical Phenomena", Knapp, Børgesen and Zuhr Eds., (North-Holland), Mat. Res. Soc. Symposium Proceedings, Vol. 157, p. 461, 1990

2.104. P.W. Mertens, J. Leclair, H.E. Maes and W. Vandervorst, J. Appl. Phys., Vol. 67, p. 7337, 1990

2.105. J. Knapp, L.P. Allen, and P.M. Zavaracky, Proc. of the IEEE SOS/SOI Technology Conference, p. 80, 1989

2.106. C. Karulkar, R.J. Hillard, and P. Rai-Choudhury, Proc. of the IEEE SOS/SOI Technology Conference, p. 97, 1989

2.107. G.K. Celler and L.E. Trimble, in "Energy Beam-Solid Interactions and Transient Thermal Processing", Fan and Johnson Eds., (North-Holland), MRS Symposium Proceedings, Vol. 23, p. 567, 1984

2.108. L. Jastrzebski, J.F. Corboy, J.T. McGinn, and R. Pagliaro, Jr., J. Electrochem Soc., Vol. 130, p. 1571, 1983

2.109. L. Jastrzebski and A.G. Kokkas, in "Energy Beam-Solid Interactions and Transient Thermal Processing", Fan and Johnson Eds., (North-Holland), MRS Symposium Proceedings, Vol. 23, p. 417, 1984

2.110. L. Jastrzebski, J. Crystal Growth, Vol. 63, p. 493, 1983

2.111. R.P. Zingg, G.W. Neudeck, B. Höfflinger, and S.T. Liu, J. Electrochem. Soc., Vol. 133, p.1274, 1986

2.112. S.T. Liu, P. Fechner, J. Friedrick, G. Neudeck, L. Velo, L. Bousse, and J. Plummer, Proc. IEEE SOS/SOI Technology Workshop, p. 16, 1988

2.113. D. R. Bradbury, T.I. Kamins, and C.W. Tsao, J. Appl. Phys., Vol. 55, p. 519, 1984

2.114. T.I. Kamins and D. R. Bradbury, IEEE Electron Device Lett., Vol. 5, p. 449, 1984

2.115. R.P. Zingg, B. Höfflinger, and G.W. Neudeck, IEDM Technical Digest, p. 909, 1989

2.116. R.P. Zingg, H.G. Graf, W. Appel, P. Vöhringer, and B. Höfflinger, Proc. IEEE SOS/SOI Technology Workshop, p. 52, 1988

2.117. A. Ogura and Y. Fujimoto, Ext. Abstracts of 8th Internat. Workshop on Future Electron Devices, Kochi, Japan, p. 73, 1990, and
A. Ogura and Y. Fujimoto, Appl. Phys. Lett., Vol. 55, p. 2205, 1989

2.118. P.J. Schubert and G.W. Neudeck, IEEE Electron Device Letters, Vol. 11, p. 181, 1990

2.119. Y.Kunii, M. Tabe, and K. Kajiyama, J. Appl. Phys., Vol. 54, p. 2847, 1983, and Y.Kunii, M. Tabe, and K. Kajiyama, J. Appl. Phys., Vol. 56, p. 279, 1984

2.120. H. Ishiwara, H. Yamamoto, S. Furukawa, M. Tamura, and T. Tokuyama, Appl. Phys. Lett., Vol. 43, p. 1028, 1983

2.121. J.A Roth, G.L. Olson, and L.D. Hess, in "Energy Beam-Solid Interactions and Transient Thermal Processing", Fan and Johnson Eds., (North-Holland), MRS Symposium Proceedings, Vol. 23, p. 431, 1984

2.122. H. Ishiwara, H. Yamamoto, and S. Furukawa, Ext. Abstracts of 2nd Internat. Workshop on Future Electron Devices, Shuzenji, Japan, p. 63, 1985

2.123. Y. Kunii and M. Tabe, Ext. Abstracts of 2nd Internat. Workshop on Future Electron Devices, Shuzenji, Japan, p. 69, 1985

2.124. T. Dan, H. Ishiwara, and S. Furukawa, Ext. Abstracts of 5th Internat. Workshop on Future Electron Devices, Miyagi-Zao, Japan, p. 189, 1988

2.125. M. Miyao, M. Moniwa, K. Kusukawa, and S. Furukawa, J. Appl. Phys., Vol. 64, p. 3018, 1988

2.126. K. Imai, Solid-State Electron., Vol. 24, p.59, 1981

2.127. S.S. Tsao, IEEE Circuits and Devices Magazine, Vol. 3, p. 3, 1987

2.128. K. Ansai, F. Otoi, M. Ohnishi, and H. Kitabayashi, Proc. IEDM, p. 796, 1984

2.129. S. Muramoto, H. Unno, and K. Ehara, Proc. Electrochem. Soc. Meeting, Boston, Vol. 86, p. 124, May 1986

2.130. L. Nesbit, Tech. Digest of IEDM, p. 800, 1984

2.131. R.P. Holmstrom and J.Y. Chi, Appl. Phys. Lett., Vol. 42, p. 386, 1983

2.132. K. Barla, G. Bomchil, R. Herino, and. A. Monroy, IEEE Circuits and Devices Magazine, Vol. 3, p. 11, 1987

2.133. E.J. Zorinsky, D. B. Spratt, and R. L. Vinkus, Tech. Digest IEDM, p. 431, 1986

2.134. K. Barla, G. Bomchil, R. Herino, J.C. Pfister, and J. Baruchel, J. Cryst. Growth, Vol. 68, p. 721, 1984, and
K. Barla, R. Herino, and G. Bomchil, J. Appl. Phys, Vol. 59, p. 439, 1986

2.136. M.I.J. Beale, N.G. Chew, A.G. Cullis, D.B. Garson, R.W. Hardeman, D.J. Robbins, and I.M. Young, J. Vac. Sci. Technol. B, p. 732, 1985

2.137. H. Takai and T. Itoh, J. Electronic Materials, Vol. 12, p.293, 1983

2.138. T.Mano, T. Baba, H. Sawada, and K. Imai, Tech. Digest Symposium on VLSI Technology, p. 12, 1982

2.139. K. Ehara, H. Unno, and S. Muramoto, Electrochem. Soc. Extended Abstracts, Vol. 85-2, p. 457, 1985

2.140. N.J. Thomas, J.R. Davis, K.J. Reeson, P.L.F. Hemment, J. Keen, J. Castledine, D. Brumhead, M. Goulding, J. Alderman, J.P.G. Farr, and L.G. Earwaker, Proc. IEEE SOS/SOI Technology Workshop, p. 39, 1988
2.140. M.Watanabe and A. Tooi, Japan. J. Appl. Phys., Vol. 5, p. 737, 1966

2.141. K. Izumi, M. Doken, and H. Ariyoshi, Electronics Letters, Vol. 14, p. 593, 1978

2.142. K. Izumi, Y. Omura, M. Ishikawa, and E. Sano, Techn. Digest of the Symposium on VLSI Technology, p. 10, 1982

2.143. J. Stoemenos, C. Jaussaud, M. Bruel, and J. Margail, J. Crystal Growth, Vol. 73, p.546, 1985

2.144. G.K. Celler, P.L.F. Hemment, K.W. West, and J.M. Gibson, Appl. Phys. Lett., Vol.48, p. 532, 1986

2.145. J.R. Davis, A. Robinson, K. Reeson, and P.L.F. Hemment, Proc. IEEE SOS/SOI Technology Workshop, p. 71, 1987

2.146. J.P. Colinge and T.I. Kamins, Proc. IEEE SOS/SOI Technology Workshop, p. 69, 1987

2.147. D. Hill, P. Fraundorf, and G. Fraundorf, Proc. IEEE SOS/SOI Technology Workshop, p. 29, 1987, and
D. Hill, P. Fraundorf, and G. Fraundorf, J. Appl. Phys., Vol. 63, p. 4933, 1988

2.148. J.P. Colinge, J. Kang, W. McFarland, C. Stout, and R. Walker, Proc. IEEE SOS/SOI Technology Workshop, p. 68, 1988

2.149. W.A. Krull and J.C. Lee, Proc. IEEE SOS/SOI Technology Workshop, p. 69, 1988

2.150. F. Namavar, E. Cortesi, B. Buchanan, and P. Sioshansi, Proc. IEEE SOS/SOI Technology Workshop, p. 117, 1989

2.151. T. W. Houston, H. Lu, P. Mei, T.G.W. Blake, L.R. Hite, R. Sundaresan, M. Matloubian, W.E. Bailey, J. Liu, A. Peterson, and G. Pollack, Proc. IEEE SOS/SOI Technology Workshop, p. 137, 1989

2.152. W.F. Kraus and J.C. Lee, Proc. IEEE SOS/SOI Technology Workshop, p. 173, 1989

2.153. A.J. Auberton-Hervé, B. Giffard, and M. Bruel, Proc. IEEE SOS/SOI Technology Workshop, p. 169, 1989

2.154. G.K. Celler, A. Kamgar, H.I. Cong, R.L. Field, S.J. Hillenius, W.S. Lindenberger, L.E. Trimble, and J.C. Sturm, Proc. IEEE SOS/SOI Technology Workshop, p. 139, 1989

2.155. H. Miki, Y. Omura, T. Ohmameuda, M. Kumon, K. Asada, K. Izumi, T. Asai, and T. Sugano, Techn. Digest of IEEE Internat. Electron Device Meeting (IEDM), p. 906, 1989

2.156. M.H. Badawi and K.V. Anand, J. Phys. D, Vol. 10, p. 1931, 1977

2.157. E.A. Mayell-Ondruz and I.H. Wilson, Thin Solid Films, Vol. 114, p. 357, 1984

2.158. P.L.F. Hemment, in "Semiconductor-On-Insulator and Thin Film Transistor Technology", Chiang, Geis and Pfeiffer Eds., (North-Holland), MRS Symposium Proceedings, Vol. 53, p. 207, 1986

2.159. P.L.F. Hemment, K.J. Reeson, J.A. Kilner, R.J. Chater, C. Maesh, G.R. Booker, J.R. Davis, and G.K. Celler, Nucl. Instr. Meth. Phys. Res., Vol. B21, p. 129, 1987

2.160. C. Jaussaud, J. Margail, J. Stoemenos, and M. Bruel, in " Silicon-On-Insulator and Buried Metals in Semiconductors", Sturm, Chen, Pfeiffer, and Hemment Eds, (North-Holland), MRS Symposium Proceedings, Vol. 107, p. 17, 1988

2.161. L. Jastrzebski, J.F. Corboy, C.W. Magee, J.H. Thomas, A.C. Ipri, D.A. Peters, G.W. Cullen, and H. Friedman, J. Electrochem. Soc., Vol. 135, p. 1746, 1988

2.162. M.A. Guerra, Proc. of the 4th International Symposium on Silicon-on-Insulator Technology and Devices, Ed. by D. Schmidt, the Electrochemical Society, Vol. 90-6, p. 21, 1990

2.163. G.K. Celler, presented at the 6th International Symposium on Silicon Materials Science and Technology, Semiconductor Silicon, Ed. by H. Huff, K. Barraclough and J.-I. Chikawa, the Electrochemical Society, Vol. 90-7, p. 472, 1990

2.164. R.S. Hockett and R.G. Wilson, Proc. of the 4th International Symposium on Silicon-on-Insulator Technology and Devices, Ed. by D. Schmidt, the Electrochemical Society, Vol. 90-6, p. 154, 1990

2.165. L. Nesbit, S. Stiffler, G. Slusser, and H. Vinton, J. Electrochem. Soc., Vol. 132, p. 2713, 1985.

2.166. E. Sobeslavsky and W. Skorupa, Phys. Stat. Sol. (A), Vol. 144, p. 135, 1989

2.167. W. Skorupa, K. Wollschläger, R. Grötzschel, J. Schöneich, E Hentschel, R. Kotte, F. Stary, H. Bartsch, and G. Götz, Nucl. Instr. and Meth. in Phys. Research, Vol. B32, p. 440, 1988

2.168. G. Zimmer and H. Vogt, IEEE Trans. on Electron Devices, Vol. 30, p. 1515, 1983

2.169. L. Nesbit, G. Slusser, R. Frenette, and R. Halbach, J. Electrochem. Soc., Vol. 133, p. 1186, 1986

2.170. W.P. Maszara, Proc. of the 4th International Symposium on Silicon-on-Insulator Technology and Devices, Ed. by D. Schmidt, the Electrochemical Society, Vol. 90-6, p. 199, 1990

2.171. W.P. Maszara, G. Goetz, A. Caviglia, and J.B. McKitterick, J. Appl. Phys., Vol. 64, p. 4943, 1988

2.172. T. Abe, M. Nakano, and T. Itoh, Proc. of the 4th International Symposium on Silicon-on-Insulator Technology and Devices, Ed. by D. Schmidt, the Electrochemical Society, Vol. 90-6, p. 61, 1990

2.173. A. Yamada, O. Okabayashi, T. Nakamura, E. Kanda, and M. Kawashima, Ext. Abstr. 5th International Workshop on Future Electron Devices, Miyagi-Zao, Japan, p. 201, 1988

2.174. N.F. Raley, Y. Sugiyama, and T. Van Duzer, J. Electrochem. Soc., Vol. 131, p. 161, 1984

2.175. H. Gotou, A. Sekiyama, T. Seki, S. Nagai, N. Suzuki, M. Hayasaka, Y. Matsukawa, M. Miyazima, Y. Kobayashi, S. Enomoto, and K. Imaoka, Techn. Digest of IEEE Internat. Electron Device Meeting (IEDM), p. 912, 1989

2.176. Proc. IEEE SOS/SOI Technology Workshop, page i, 1990

2.177. J.P. Colinge, Techn. Digest of International Electron Devices Meeting (IEDM), p. 817, 1989

2.178. B. Tillack, K. Hoeppner, H.H. Richter, and R. Banisch, Materials Science and Engineering (Elsevier Sequoia), Vol. B4, p. 237, 1989

2.179. B. Tillack, R. Banisch, H.H. Richter, K. Hoeppner, O. Joachim, J. Knopke, and U. Retzlaf, Proceedings IEEE SOS/SOI Technology Conference, p. 121, 1990

Chapter 3

3.1. S.N. Bunker, P. Sioshansi, M.M. Sanfacon, S.P. Tobin, Appl. Phys. Lett. Vol. 50, p. 1900, 1987

3.2. Z. Knittl, Optics of thin films, Wiley, New York, p. 37, 1976

3.3. F. Van de Wiele, Solid-State Imaging, NATO Advanced Study Institutes Series, Noordhoff, Leyden, p. 29, 1976

3.4. J.P. Colinge and F. Van de Wiele, J. Appl. Phys, Vol. 52, p. 4769, 1981

3.5. D.E. Aspenes, Properties of Silicon, Published by INSPEC (IEE), p. 59, 1988

3.6. T.I. Kamins and J.P. Colinge, Electronics Letters, Vol. 22, p. 1236, 1986

3.7. J. Vanhellemont, J.P. Colinge, A. De Veirman, J. Van Landuyt, W. Skorupa, M. Voelskow, and H. Bartsch, Proc. of the 4th International Symposium on Silicon-on-Insulator Technology and Devices, Ed. by D. Schmidt, the Electrochemical Society, Vol. 90-6, p. 187, 1990

3.8. H.J. Hovel, in "Semiconductors and Semimetals", Vol.11 "Solar Cells", Ed. by R.K. Wilardson and A.C. Beer, Academic Press, p. 203, 1975

3.9. P.L. Swart and B.M. Lacquet, Proceedings IEEE SOS/SOI Conference, p. 153, 1989, and
P.L. Swart and B.M. Lacquet, J. of Electronic Materials, Vol. 19,p. 809, 1990

3.10. R.M.A. Azzam and N.M. Bashara, Ellipsometry and Polarized Light, Elsevier Science Publishers, North-Holland Personal Edition, chapter 1, 1987

3.11. D.A.G. Bruggeman, Annalen der Physik, Vol. 5, p. 636, 1935

3.12. J. Whitfield and S. Thomas, IEEE Electron Device Letters, Vol. 7, p. 347, 1986

3.13. D.C. Joy, D.E. Newbury, and D.L. Davidson, J. Appl. Phys., Vol. 53, p. R81, 1982

3.14. K.A. Bezjian, H.I. Smith, J.M. Carter, and M.W. Geis, J. Electrochem. Soc, Vol. 129, p. 1848, 1982

3.15. W.K. Chu, J.W. Mayer, and M.A. Nicolet, Backscattering Spectrometry, Academic Press, N.Y., 1978

3.16. M.T. Duffy, J.F. Corboy, G.W. Cullen, R.T. Smith, R.A. Soltis, G. Harbeke, J.R. Sandercock, and M. Blumenfeld, J. Crystal Growth, Vol. 58, p. 10, 1982

3.17. G. Harbeke and L. Jastrzebski, J. Electrochem. Society, Vol. 137, p. 696, 1990

3.18. W.C. Dash, J. Appl. Phys, Vol. 27, p. 1993, 1956

3.19. D.G. Schimmel, J. Electrochem. Soc, vol 126, p. 479, 1979

3.20. F. Secco d'Aragona, J. Electrochem. Soc., Vol. 119, p. 948, 1972

3.21. E. Sirtl and A. Adler, Zeitung für Metallkunde, Vol. 52, p. 529, 1961

3.22. M. Wright Jenkins, J. Electrochem. Soc., Vol. 124, p. 757, 1977

3.23. T.R. Guilinger, M.J. Kelly, J.W. Medernach, S.S. Tsao, J.O. Steveson, and H.D.T. Jones, Proc. IEEE SOS/SOI Technology Conference, p. 93, 1989, and
M.J. Kelly, T.R. Guilinger, J.W. Medernach, S.S. Tsao, H.D.T. Jones, and J.O. Steveson, Proceedings of the fourth international Symposium on Silicon-on-Insulator Technology and Devices, ed. by D.N. Schmidt, Vol. 90-6, The Electrochemical Society, p. 120, 1990

3.24. K.K. Ng, G.K. Celler, E.J. Povilonis, R.C. Frye, H.J. Leamy, and S.M. Sze, IEEE Electron Device Letters, Vol. 2, p. 316, 1981

3.25. J.P. Colinge, H. Morel, and J.P. Chante, IEEE Trans. on Electron Devices, Vol. 30, p. 197, 1983

3.26. T. Nishimura, K. Sugahara, S. Kusunoki, and Y. Akasaka, Ext. Abstracts of the 17th Conference of on Solid-State Devices and Materials, Tokyo, p. 1147, 1985

3.27. T.I. Kamins, Electronics Letters, Vol. 23, p. 175, 1987

3.28. L. Jastrzebski, J.T. McGinn, P. Zanzucchi, and B. Cords, J. Electrochem. Soc., Vol. 137, p. 306, 1990

3.29. S. Cristoloveanu, J. Pumfrey, E. Scheid, P.L.F. Hemment, and R.P. Arrowsmith, Electronics Letters, Vol. 21, p. 802, 1985

3.30 T.S. Moss, Optical Properties of Semiconductors, Butterworths, London, Chapter 4, 1959

3.31. A.M. Goodman, J. Appl. Phys, Vol. 53, p. 7561, 1982

3.32. L. Jastrzebski, G. Cullen, and R. Soydan, J. Electrochem. Society, Vol. 137, p. 303, 1990

3.33. M.A. Guerra, Proc. of the 4th International Symposium on Silicon-on-Insulator Technology and Devices, Ed. by D. Schmidt, the Electrochemical Society, Vol. 90-6, p. 21, 1990

3.34. H.S. Chen, F.T. Brady, S.S. Li, and W.A. Krull, IEEE Electron Device Letters, Vol. 10, p. 496, 1989

3.35. H.S. Chen and S.S. Li, Proc. of the 4th International Symposium on Silicon-on-Insulator Technology and Devices, Ed. by D. Schmidt, the Electrochemical Society, Vol. 90-6, p. 328, 1990

3.36. D.P. Vu and J.C. Pfister, Appl. Phys. Letters, Vol. 47, p. 950, 1985

3.37. T. Elewa, H. Haddara, and S. Cristoloveanu, in "Solid-State Devices", Ed. By. G. Soncini and P.U. Calzolari, Elsevier Science Publishers (North-Holland), p. 599, 1988

3.38. M. Zerbst, Z. Angew. Phys., Vol. 22, p. 30, 1966

3.39. P.K. McLarty, T. Elewa, B. Mazhari, M. Mukherjee, T. Ouisse, S. Cristoloveanu, D.E. Ioannou, and D.P. Vu, Proceedings IEEE SOS/SOI Technology Conference, p. 54, 1989

3.40. T. Elewa, Ph.D. Thesis, ENSERG-LPCS, Grenoble (France), p. 90, July 1990

3.41. M. Haond and J.P. Colinge, Electronics Letters, Vol. 25, p.1640 , 1989

3.42. D. Flandre and F. Van De Wiele, IEEE Electron Device Letters, Vol. 9, p. 296, 1988.

3.43. M. Gaitan and P. Roitman, Proceedings IEEE SOS/SOI Technology Conference, p. 48, 1989

3.44. J.H. Lee and S. Cristoloveanu, IEEE Electron Device Letters, Vol. 7, p. 537, 1986.

3.45 J.S. Brugler and P.G.A. Jespers, IEEE Trans. Electron Devices, Vol. 16, p. 297, 1969

3.46. G. Groeseneken, H.E. Maes, N. Beltran, and R.F. Dekeersmaecker, IEEE Trans. Electron Devices, Vol. 31, p. 42, 1984

3.47. T. Elewa, H. Haddara, S. Cristoloveanu and M. Bruel, J. de Physique, Vol. 49, no 9-C4, p. C4-137, 1988

Chapter 4

4.1. T. Elewa, B. Kleveland, B. Boukriss, T. Ouisse, A. Chovet, S. Cristoloveanu, and J. Davis, Proceedings IEEE SOS/SOI Technology Conference, p. 35, 1989

4.2. R.K. Smeltzer and J.T. McGinn, Proceedings IEEE SOS/SOI Technology Workshop, p. 32, 1987

4.3. S.S. Tsao, D.M. Fletwood, V. Kaushik, A.K. Datye, L. Pfeiffer, and G.K. Celler, Proceedings IEEE SOS/SOI Technology Workshop, p. 33, 1987

4.4. R.B. Marcus and T.T. Sheng, J. Electrochem. Soc., Vol. 129, p. 1278, 1982

4.5. M. Matloubian, R. Sundaresan, and H. Lu, Proceedings IEEE SOS/SOI Technology Workshop, p. 80, 1988

4.6. M. Haond and O. Le Néel, Proceedings IEEE SOS/SOI Technology Conference, p. 132, 1990, and
O. Le Néel, M.D. Bruni, J. Galvier, and M. Haond, in "ESSDERC 90", Adam Hilger Publisher, Ed. by. W. Eccleston and P.J. Rosser, p. 13, 1990

4.7. M. Haond, O. Le Neel, G. Mascarin, and J.P. Gonchond, Proceedings IEEE SOS/SOI Technology Conference, p. 68, 1989, and
M. Haond, O. Le Neel, G. Mascarin, and J.P. Gonchond, in ESSDERC'89, European Solid-State Device Research Conference, Berlin, Ed. by. A. Heuberger, H. Ryssel and P. Lang, Springer-Verlag, p. 893, 1989

4.8. T. Aoki, M. Tomizawa, and A. Yoshii, IEEE Trans. on Electron Devices, Vol. 36, p. 1725, 1989

4.9. J.P. Colinge, Proceedings IEEE SOS/SOI Technology Conference, p. 13, 1989

4.10. T. Nishimura, Y. Yamaguchi, H. Miyatake, and Y. Akasaka, Proceedings IEEE SOS/SOI Technology Conference, p. 132, 1989

4.11. N.K. Annamalai and M.C. Biwer, IEEE Trans. Nuclear Science, Vol. 35, p. 1372, 1988

4.12. Y. Omura and K. Izumi, IEEE Trans. on Electron Devices, Vol. 35, p. 1391, 1988

4.13. M. Haond, in ESSDERC'89, European Solid-State Device Research Conference, Berlin, Ed. by. A. Heuberger, H. Ryssel and P. Lang, Springer-Verlag, p. 881, 1989

Chapter 5

5.1. D. Flandre and F. Van de Wiele, Proc. of the IEEE SOS/SOI Technology Conference, p. 27, 1989

5.2. N.J. Thomas and J.R. Davis, Proc. of the IEEE SOS/SOI Technology Conference, p. 130, 1989

5.3. see, for example, S.M. Sze, Physics of Semiconductor Devices, 2nd Ed., New York: J. Wiley & Sons, 1981

5.4. H.K. Lim and J.G. Fossum, IEEE Trans. on Electron Devices, Vol. 30, p. 1244, 1983

5.5. J.P. Colinge, IEEE Electron Device Lett., Vol. 6, p. 573, 1985

5.6. R.S. Muller and T.I. Kamins, Device Electronics for Integrated Circuits, J. Wiley & Sons, p. 436, 1986

5.7. J. Witfield and S. Thomas, IEEE Electron Dev. Lett., Vol. 7, p. 347, 1986

5.8. J.P. Colinge, Microelectronic Engineering, Vol. 8, p. 127, 1988

5.9. R.S. Muller and T.I. Kamins, Device Electronics for Integrated Circuits, J. Wiley & Sons, p. 487, 1986

5.10. S.Veeraraghavan and J.G. Fossum, IEEE Trans. on Electron Devices, Vol. 35, p. 1866, 1988

5.11. T.W. MacElwee and D.I. Calder, Proceedings of the second international Symposium on Ultra Large Scale Integration Science and Technology, Ed. by C.M. Osburn and J.M. Andrews, Vol. 89-9, The Electrochemical Society, p. 693, 1989, and
T.W. MacElwee and I.D. Calder, Proceedings IEEE SOS/SOI Technology Conference, p. 171, 1989

5.12. J.P. Colinge, Techn. Digest of International Electron Devices Meeting (IEDM), p. 817, 1989

5.13. J.G. Fossum, Proceedings of the fourth international Symposium on Silicon-on-Insulator Technology and Devices, ed. by D.N. Schmidt, Vol. 90-6, The Electrochemical Society, p. 491, 1990

5.14. T. Sekigawa and Y. Hayashi, Solid-State Electron., Vol. 27, pp. 827, 1984

5.15. D. Hisamoto, T. Kaga, Y. Kawamoto, and E. Takeda, Techn. Digest of International Electron Devices Meeting (IEDM), p. 833, 1989

5.16. S.Veeraraghavan and J.G. Fossum, IEEE Trans. on Electron Devices, Vol. 36, p. 522, 1989

5.17. H.K. Lim and J.G. Fossum, IEEE Trans. on Electron Devices, Vol. 31, p. 401, 1984

5.18 H.K. Lim and J.G. Fossum, IEEE Trans. on Electron Devices, Vol. 32, p. 446, 1985

5.19. R.J. Van Overstraeten, G.J. Declerck, and P.A. Muls, IEEE Trans. Electron Devices, Vol. ED-20, p. 1150, 1973, and
S.M. Sze, Physics of Semiconductor Devices, Wiley & Sons, p. 446, 1981

5.20. D.J. Wouters, J.P. Colinge, and H.E. Maes, IEEE Trans. Electron Devices, Vol. ED-37, p. 2022, 1990

5.21. J.P. Colinge, IEEE Electron Device Lett., Vol. EDL-7, p. 244, 1986

5.22. J.P. Colinge, Ext. Abstracts of 5th Internat. Workshop on Future Electron Devices, Miyagi-Zao, Japan, p. 105, 1988

5.23. K.Asada, H. Miki, M. Kumon, and T. Sugano, Ext. Abstracts of 8th Internat. Workshop on Future Electron Devices, Kochi, Japan, p. 165, 1990

5.24. S.C. Sun and J.D. Plummer, IEEE Trans. on Electron Devices, Vol. 27, pp. 1497, 1980

5.25. M. Yoshimi, H. Hazama, M. Takahashi, S. Kambayashi, T. Wada, K. Kato, and H. Tango, IEEE Trans. on Electron Devices, Vol. 36, pp. 493, 1989

5.26. A. Yoshino, Proceedings of the fourth international Symposium on Silicon-on-Insulator Technology and Devices, ed. by D.N. Schmidt, Vol. 90-6, The Electrochemical Society, p. 544, 1990

5.27. F. Balestra, S. Cristoloveanu, M. Benachir, J. Brini, and T. Elewa, IEEE Electron Device Lett., Vol. 8, pp. 410, 1987

5.28. J.C. Sturm, Silicon-on-Insulator and Buried Metals in Semiconductors, MRS Symp. Proc., Ed. by J.C. Sturm, C.K. Chen, L. Pfeiffer, and P.L.F. Hemment, Vol.107, p. 295, 1988, and
J.C. Sturm, K. Tokunaga, and J.P. Colinge, IEEE Electron Device Letters, Vol. 9, p. 460, 1988

5.29. J.C. Sturm and K. Tokunaga, Electronics Letters, Vol. 25, p. 1233, 1989

5.30. J.G. Fossum, J.Y. Choi, and R. Sundaresan, IEEE Trans. on Electron Devices, Vol. 37, p. 724, 1990

5.31. P. Antognetti and G. Massobrio, Semiconductor Device Modeling with Spice, McGraw-Hill, p. 185, 1988

5.32. B. Dierickx, L. Warmerdam, E. Simoen, J. Vermeiren, and C. Claeys, IEEE Trans. on Electron Devices, Vol. 35, p. 1120, 1988

5.33. J. Tihanyi and H Schlötterer, IEEE Trans. on. Electron Devices, Vol. 22, p. 1017, 1975

5.34. G. Merckel, Nato Course on Process and Device Modeling for Integrated Circuit Design, Ed. by F. Van de Wiele, W. Engl and P. Jespers, Groningen, The Netherlands, Noordhoff, p. 725, 1977

5.35. J.P. Colinge, IEEE Electron Device Lett., Vol. 9, p. 97, 1988

5.36. K.M. Cham, S.Y. Oh, D. Chin, and J.L. Moll, Computer Aided Design and VLSI Device Development, Hingham, MA, Kluwer Academic Publishers, p. 240, 1986

5.37. P.K. Ko, S. Tam, C. Hu, S.S. Wong, and C.G. Sodini, Techn. Digest of International Electron Devices Meeting (IEDM), p. 88, 1984

5.38. C. Hu, Techn. Digest of International Electron Devices Meeting (IEDM), p. 176, 1983

5.39. C.Hu, S.C. Tam, F.C. Hsu, P.K. Ko, T.Y. Chan, and T.W. Terrill, IEEE Trans. on. Electron Devices, Vol. 32, p. 375, 1985

5.40. J.P. Colinge, IEEE Trans. on Electron Devices, Vol. 34, p. 2173, 1987

5.41. J.R. Davis, A.E. Glaccum, K. Reeson, and P.L.F. Hemment, IEEE Electron Device Letters, Vol. 7, p. 570, 1986

5.42. J.Y. Choi and J.G. Fossum, Proceedings IEEE SOS/SOI Technology Conference, p. 21, 1990

5.43. C.E.D. Chen, M. Matloubian, R. Sundaresan, B.Y. Mao, C.C. Wei, and G.P. Pollack, IEEE Electron Device Letters, Vol. 9, p. 636, 1988

5.44. R. Sundaresan and C.E.D. Chen, Proceedings of the fourth international Symposium on Silicon-on-Insulator Technology and Devices, ed. by D.N. Schmidt, Vol. 90-6, The Electrochemical Society, p. 455, 1990

5.45. A.J. Auberton-Hervé, Proceedings of the fourth international Symposium on Silicon-on-Insulator Technology and Devices, ed. by D.N. Schmidt, Vol. 90-6, The Electrochemical Society, p. 544, 1990

5.46. H.S. Sheng, S.S. Li, R.M. Fox, and W.S. Krull, IEEE Trans. on Electron Devices, Vol. 36, no. 3, p. 488, 1989

5.47. Grove, A.S., Physics and Technology of Semiconductor Devices, J. Wiley & Sons, pp. 230, 1967

5.48. M. Haond and J.P. Colinge, Electronics Letters, Vol. 25, p.1640 , 1989

5.49. K.K. Young and J.A. Burns, IEEE Transactions on Electron Devices, Vol. 35, p. 426, 1988

5.50. J.P. Colinge, IEEE Trans. on Electron Devices, Vol. 37, p. 718, 1990

5.51. D.P. Vu, A. Chantre, D. Ronzani, and J.C. Pfister, in "Semiconductor-On-Insulator and Thin Film Transistor Technology", Chiang, Geis and Pfeiffer Eds., (North-Holland), MRS Symposium Proceedings, Vol. 53, p. 357, 1986

5.52. Grove, A.S., Physics and Technology of Semiconductor Devices, J. Wiley & Sons, pp. 326, 1967

5.53. J.G. Fossum, R. Sundaresan, and M. Matloubian, IEEE Transactions on Electron Devices, Vol. 8, p. 544, 1987

Chapter 6

6.1. S. Malhi, C.C. Shen, M. Anderson, K. Bean, D. Yeakley, G. Gopffarth, R. Sundaresan, K. Lindbergh, and J. Smith, Extended abstracts of the ECS Spring meeting, p. 427, 1989

6.2. S. Gaul, J. Delgado, G. Rouse, and C. McLachlan, Proceedings of the IEEE SOS/SOI Workshop, p. 72, 1987

6.3. C. Cahill and A. Mathewson, Extended abstracts of the ECS Spring meeting, p. 429, 1989

6.4. J.P. Colinge and S.Y. Chiang, IEEE Electron Device Letters, Vol. 7, p. 697, 1986

6.5. L. Hobbs, A. Mathewson, and W.A. Lane, Extended abstracts of the European SOI Workshop, Ed. by Bensahel and Bomchil, p. F-09, 1988

6.6. T. Ohno, S. Matsumoto, and K. Izumi, Electronics Letters, Vol. 25, p. 1071, 1989

6.7. J.P. Colinge, IEEE Transactions on Electron Devices, Vol. 33, p. 203, 1986

6.8. Y. Omura, Japanese Journal of Applied Physics, Vol. 22, Suppl. 22-1, p. 263, 1983

6.9. J.P. Colinge, IEEE Transactions on Electron Devices, Vol. 34, p. 845, 1987

6.10. S. Verdonck-Vandenbroek, S.S. Wong, and P.K. Ko, Technical Digest of IEDM, p 406, 1988

6.11. J.P. Colinge, Electronics Letters, Vol. 23, p. 1023, 1987

6.12. D. Hisamoto, T. Kaga, Y. Kawamoto, and E. Takeda, Technical Digest of IEDM, p. 833, 1989

6.13. J.P. Colinge, M.H. Gao, A. Romano, H. Maes, and C. Claeys, Proceedings IEEE SOS/SOI Technology Conf., p. 137, 1990

6.14. T. Sekigawa and Y. Hayashi, Solid-State Electronics, Vol. 27, p. 827, 1984

6.15. F. Balestra, M. Benachir, J. Brini, L. Sweid, G. Ghibaudo, and N. Guillemot, Proceedings IEEE SOS/SOI Technology Conference, p. 25, 1989

6.16. F. Balestra, S. Cristoloveanu, M. Benachir, J. Brini, and T. Elewa, IEEE Electron Device Letters, Vol. 8, p. 410, 1987

6.17. D.C. Mayer, IEEE Trans. on Electron Devices, Vol. 37, p. 1280, 1990

6.18. J.P. Colinge, Electronics Letters, Vol. 22, p. 886, 1986

6.19. J.C. Sturm, J.P. McVittie, J.F. Gibbons, and L. Pfeiffer, IEEE Electron Device Letters, Vol. 8, p. 104, 1987

6.20. J.C. Sturm, Ph.D. thesis, Stanford University, p. 95, 1985, and J.C. Sturm and J.F. Gibbons, in "Semiconductor-On-Insulator and Thin Film Transistor Technology", Chiang, Geis and Pfeiffer Eds., (North-Holland), MRS Symposium Proceedings, Vol. 53, p. 395, 1986

6.21. S.J. Gaul, J.A. Delgado, G.V. Rouse, C.J. McLachlan, and W.A. Krull, Proceedings IEEE SOS/SOI Technology Conference, p. 101, 1989

6.22. M.D. Church, Proceedings IEEE SOS/SOI Technology Conference, p. 175, 1989

6.23. D.G. Platteter and T.F. Cheek, Jr., IEEE Trans. Nuclear Science, Vol. 35, p. 1350, 1988

6.24. J.P. Blanc, J. Bonaime, E. Delevoye, J. Gauthier, J. de Pontcharra, R. Truche, E. Dupont-Nivet, J.L. Martin, and J. Montaron, Proceedings IEEE SOS/SOI Technology Conference, p. 85, 1990

6.25. X. Xiao, J.C. Sturm, P.V. Schwartz, and K.K. Goel, Proceedings IEEE SOS/SOI Technology Conference, p. 171, 1990

Chapter 7

7.1. G.C. Messenger and M.S. Ash, The Effects of Radiation on Electronic Systems, Van Nostrand Reinhold Company, New York, 1986

7.2. G.C. Messenger and M.S. Ash, The Effects of Radiation on Electronic Systems, Van Nostrand Reinhold Company, New York, 1986, p. 307

7.3. J.L. Leray, Microelectronics Engineering, Vol. 8, p. 187, 1988

7.4. G.E.Davis, L.R. Hite, T.G.W. Blake, C.E. Chen, H.W. Lam, R. DeMoyer, IEEE Trans. on Nuclear Science, Vol. 32, p. 4432, 1985

7.5. J.L. Leray, E. Dupont-Nivet, O. Musseau, Y.M. Coïc, A. Umbert, P. Lalande, J.F. Péré, A.J. Auberton-Hervé, M. Bruel, C. Jaussaud, J. Margail, B. Giffard, R. Truche, and F. Martin, IEEE Trans. on Nuclear Science, Vol. 35, p. 1355, 1988

7.6. G.E. Davis, in "Silicon-On-Insulator and Buried Metals in Semiconductors", Sturm, Chen, Pfeiffer and Hemment Eds., (North-Holland), MRS Symposium Proceedings, Vol. 107, p. 317, 1988

7.7. L.W. Massengill, D.V. Kerns, Jr., S.E. Kerns, and M.L. Alles, IEEE Electron Device Letters, Vol. 11, p. 98, 1990

7.8. F. Wulf, D. Braunig, and A. Boden, ECFA STUDY WEEK on Instrumentation Technology for High-Luminosity Hadron Colliders, Ed. by E. Fernands and G. Jarlskog, Proc. Vol. 1, p. 109, 1989

7.9. J.L. Leray, E. Dupont-Nivet, J.F. Péré, O. Musseau, P. Lalande, and A. Umbert, Proceedings SOS/SOI Technology Workshop, p. 114, 1989

7.10. G.C. Messenger and M.S. Ash, The Effects of Radiation on Electronic Systems, Van Nostrand Reinhold Company, New York, 1986, p. 243

7.11. D.M. Fletwood, S.S. Tsao, and P.S. Winokur, IEEE Trans. on Nuclear Science, Vol. 35, p. 1361, 1988

7.12. D.C. Mayer, IEEE Trans. on Electron Devices, Vol. 37, p. 1280, 1990

7.13. A.C. Ipri, L. Jastrzebski, and D. Peters, IEEE Electron Device Letters, Vol. 10, p. 571, 1989

7.14. G.C. Messenger and M.S. Ash, The Effects of Radiation on Electronic Systems, Van Nostrand Reinhold Company, New York, 1986, p. 267

7.15. W.A. Krull and J.C. Lee, Proceedings IEEE SOS/SOI Technology Workshop, p. 69, 1988

7.16. W.P. Maszara, Proc. of the 4th International Symposium on Silicon-on-Insulator Technology and Devices, Ed. by D. Schmidt, the Electrochemical Society, Vol. 90-6, p. 199, 1990

7.17. D.P. Vu, M.J. Boden, W.R. Henderson, N.K. Cheong, P.M. Zavaracky, D.A. Adams, and M.M. Austin, Proceedings IEEE SOS/SOI Technology Conference, p. 165, 1989

7.18. G. Groeseneken, J.P. Colinge, H.E. Maes, J.C. Alderman and S. Holt, IEEE Electron Device Letters, Vol. 11, p. 329, 1990

7.19. R.S. Muller and T.I. Kamins, Device Electronics for Integrated Circuits, 2nd Edition, J. Wiley & Sons Eds, p. 56, 1986

7.20. S.M. Sze, Physics of Semiconductor Devices, 2nd Edition, J. Wiley & Sons Eds, p. 91, 1981

7.21. T.W. Houston, H. Lu, P. Mei, T.G.W. Blake, L.R. Hite, R. Sundaresan, M. Matloubian, W.E. Bailey, J. Liu, A. Peterson, and G. Pollack, Proceedings IEEE SOS/SOI Technology Conference, p. 137, 1989

7.22. K. Terrill, J. Woo, and P.K. Vasudev, Ext. Abstr. of the Internat. Electron Device Meeting (IEDM), p. 294, 1988

7.23. N. Kistler, J. Woo, K. Terrill, and P.K. Vasudev, Proceedings IEEE SOS/SOI Technology Conference, p. 56, 1989

7.24. K.K. Young and B.Y. Tsaur, IEEE Electron Device Letters, Vol. 11, p. 126, 1990

7.25. T. Elewa, F. Balestra, S. Cristoloveanu, I.M. Hafez, J.P. Colinge, A.J. Auberton-Hervé, and J.R. Davis, IEEE Trans. on Electron Devices, Vol. 37, p. 1007, 1990

7.26. B. Dierickx, L. Warmerdam, E. Simoen, J. Vermeiren, and C. Claeys, IEEE Trans. Electron Devices, Vol. 35, p. 1120, 1988

Chapter 8

8.1. D.M. Fletwood, F.V. Thome, S.S. Tsao, P.V. Dressendorfer, V.J. Dandini, and J.R. Schwank, IEEE Trans. Nucl. Sci, Vol. 35, p. 1099, 1988

8.2. G.E.Davis, L.R. Hite, T.G.W. Blake, C.E. Chen, H.W. Lam, R. DeMoyer, IEEE Trans. on Nuclear Science, Vol. 32, p. 4432, 1985

8.3. W.F. Kraus and J.C. Lee, Proceedings SOS/SOI Technology Conference, p. 173, 1989

8.4. J.L. Leray, E. Dupont-Nivet, M Raffaelli, Y.M. Coïc, O. Musseau, J.F. Péré, P. Lalande, J. Brédy, A.J. Auberton-Hervé, M. Bruel, and B. Giffard, Annales de Physique, Colloque n°2, Vol. 14, suppl. to n°6, p. 565, 1989, and
J.L. Leray, E. Dupont-Nivet, J.F. Péré, Y.M. Coïc, A.J. Auberton-Hervé, M. Bruel, and B. Giffard, Tech. Prog. of IEEE Nuclear and Space Radiation Effects Conference (NSREC), paper PG-7, p. 35, 1990 and
J.L. Leray, E. Dupont-Nivet, J.F. Péré, Y.M. Coïc, M. Rafaelli, A.J. Auberton-Hervé, M. Bruel, B. Giffard and J. margail, IEEE Trans. Nucl. Sci., Vol. 37, no. 6, Dec. 1990

8.5. M.A. Guerra, Proc. of the 4th International Symposium on Silicon-on-Insulator Technology and Devices, Ed. by D. Schmidt, the Electrochemical Society, Vol. 90-6, p. 21, 1990

8.6. G.T. Goeloe, G.A. Hanidu, and K.H. Lee, Tech. Prog. of IEEE Nuclear and Space Radiation Effects Conference (NSREC), paper PG-5, p. 35, 1990

8.7. W.F. Kraus, J.C. Lee, W.H. Newman, and J.E. Clark, Tech. Prog. of IEEE Nuclear and Space Radiation Effects Conference (NSREC), paper PG-6, p. 35, 1990

8.8. C.S. Yue, J. Kueng, P. Fechner, and T. Randazzo, Proceedings SOS/SOI Technology Conference, p. 173, 1990

8.9. L.J. Palkuti, J.J. LePage, IEEE Transactions on Nuclear Science, Vol. 29, p. 1832, 1982

8.10. J.L. Leray, E. Dupont-Nivet, J.F. Péré, O. Musseau, P. Lalande, and A. Umbert, Proceedings SOS/SOI Technology Conference, p. 114, 1989

8.11. W.A. Krull and J.C. Lee, Proceedings SOS/SOI Technology Workshop, p. 69, 1989

8.12. W.P. Maszara, Proc. of the 4th International Symposium on Silicon-on-Insulator Technology and Devices, Ed. by D. Schmidt, the Electrochemical Society, Vol. 90-6, p. 199, 1990

8.13. G.A. Armstrong, W.D. French, and J.C. Alderman, Proceedings IEEE SOS/SOI Technology Conference, p. 17, 1990

8.14. L.J. McDaid, S. Hall, J.S. Marsland, W. Eccleston, J.C. Alderman, K.R. Cook, R.J. Bunyan and M.J. Uren, Proceedings IEEE SOS/SOI Technology Conference, p. 19, 1990

8.15. J.Y. Choi and J.G. Fossum, Proceedings IEEE SOS/SOI Technology Conference, p. 21, 1990

8.16. Y. Yamaguchi, T. Iwamatsu, T. Nishimura, and Y. Akasaka, Proceedings IEEE SOS/SOI Technology Conference, p. 23, 1990

8.17. L.J. McDaid, S. Hall, W. Eccleston ,P. Watkinson, and J.C. Alderman, Proceedings IEEE SOS/SOI Technology Conference, p. 141, 1990

8.18. J.P. Colinge, Technical Digest of IEDM, p. 817, 1989

8.19 A.J. Auberton-Hervé, B. Giffard, and M. Bruel, Proceedings IEEE SOS/SOI Technology Conference, p. 169, 1989

8.20. T.W. Houston, H. Lu, P. Mei, T.G.W. Blake, L.R. Hite, R. Sundaresan, M. Matloubian, W.E. Bailey, J. Liu, A. Peterson, and G. Pollack, Proceedings IEEE SOS/SOI Technology Conference, p. 137, 1989

8.21. H. Gotou, A. Sekiyama, T. Seki, S. Nagai, N. Suzuki, M. Hayasaka, Y. Matsukawa, M. Miyazima, Y. Kobayashi, S. Enomoto, and. K. Imaoka, Technical Digest of IEDM, p. 912, 1989

8.22. A.J. Auberton-Hervé, M. Bruel, C. Jaussaud, J. Margail, W. D'Hespel, J.F. Péré, A. Vitez, and A. Tissot, Proceedings European SOI Workshop, Ed. by D. Bensahel and G. Bomchil, paper F-03, 1988

8.23. H. Miki, Y. Omura, T. Ohmameuda, M. Kumon, K. Asada, K. Izumi, T. Sakai, and T. Sugano, Technical Digest of IEDM, p. 906, 1989

8.24. J.P. Colinge, J. Kang, W. McFarland, C. Stout, and R. Walker, Proceedings IEEE SOS/SOI Technology Workshop, p. 68, 1988

8.25. J.P. Colinge, J. Kang, W. McFarland, C. Stout, and R. Walker, Proceedings Hewlett-Packard Design Technology Conference, p. 356, 1988

8.26. A. Kamgar, S.J. Hillenius, H.I. Cong, R.L. Field, and J.C. Sturm, Technical Digest of IEDM, p. 829, 1989

8.27. M. Guerra, A. Wittkower, J. Stahman, and J. Schrankler, Semiconductor international, May 1990

8.28. A.J. Auberton-Hervé, Proceedings IEEE SOS/SOI Technology Conference, p. 149, 1990

8.29. J.F. Gibbons and K.F. Lee, IEEE Electron Device Letters, Vol. 1, p. 117, 1980

8.30. J.P. Colinge, E. Demoulin, and M. Lobet, IEEE Trans. on Electron Devices, Vol. 29, p. 585, 1982

8.31. R.P. Zingg and B. Höfflinger, Tech. Digest of International Electron Device Meeting (IEDM), p. 909, 1989

8.32. L. Stringa and A. Zorat, Ext. Abstracts of the 5th Internat. Workshop on Future Electron Devices - Three-Dimensional Integration, Miyagi-Zao, Japan, p. 47, 1988

8.33. T. Nishimura and Y. Akasaka, Ext. Abstracts of the 5th Internat. Workshop on Future Electron Devices - Three-Dimensional Integration, Miyagi-Zao, Japan, p. 1, 1988,and
T. Nishimura, Y. Inoue, K. Sugahara, S. Kusunoki, T. Kumamoto, S. Nagawa, M. Nakaya, Y. Horiba, and Y. Akasaka, Tech. Digest of International Electron Device Meeting (IEDM), p. 111, 1987

8.34. A. Terao and F. Van de Wiele, IEEE Circuits and Devices Magazine, Vol. 3, no 6, p. 31, 1987

8.35. R. Aibara, Y. Mitsui, and T. Ae, Ext. Abstracts of the 8th Internat. Workshop on Future Electron Devices - Three-Dimensional ICs and Nanometer Functional Devices, Kochi, Japan, p. 113, 1990

8.36. T. Ae, Ext. Abstracts of the 5th Internat. Workshop on Future Electron Devices - Three-Dimensional Integration, Miyagi-Zao, Japan, p. 55, 1988

8.37. Y. Inoue, T. Ipposhi, T. Wada, K. Ichinose, T. Nishimura, and Y. Akasaka, Proceedings Symp. VLSI Technology, p. 39, 1989

8.38. Y. Akasaka, Ext. Abstracts of the 8th Internat. Workshop on Future Electron Devices - Three-Dimensional ICs and Nanometer Functional Devices, Kochi, Japan, p. 9, 1990

8.39. T. Kunio, K. Oyama, Y. Hayashi, and M. Moritomo, Tech. Digest of International Electron Device Meeting (IEDM), p. 837, 1989

8.40. K. Yamazaki, Y.Itoh, A.Wada, and Y. Tomita, Ext. Abstracts of the 8th Internat. Workshop on Future Electron Devices - Three-Dimensional ICs and Nanometer Functional Devices, Kochi, Japan, p. 105, 1990

8.41. K. Kioi, S. Toyoyama, and M. Koba, Tech. Digest of International Electron Device Meeting (IEDM), p. 66, 1988

8.42. S. Toyoyama, K. Kioi, K. Shirakawa, T. Shinozaki, and Ohtake, Ext. Abstracts of the 8th Internat. Workshop on Future Electron Devices - Three-Dimensional ICs and Nanometer Functional Devices, Kochi, Japan, p. 109, 1990

INDEX

9 780792 391500